Studies in Spectrum Analysis

NORMAN LOCKYER

CAMBRIDGE
UNIVERSITY PRESS

CAMBRIDGE UNIVERSITY PRESS

Cambridge, New York, Melbourne, Madrid, Cape Town,
Singapore, São Paolo, Delhi, Tokyo, Mexico City

Published in the United States of America by Cambridge University Press, New York

www.cambridge.org
Information on this title: www.cambridge.org/9781108037716

© in this compilation Cambridge University Press 2011

This edition first published 1878
This digitally printed version 2011

ISBN 978-1-108-03771-6 Paperback

CAMBRIDGE LIBRARY COLLECTION

Books of enduring scholarly value

Physical Sciences

From ancient times, humans have tried to understand the workings of the world around them. The roots of modern physical science go back to the very earliest mechanical devices such as levers and rollers, the mixing of paints and dyes, and the importance of the heavenly bodies in early religious observance and navigation. The physical sciences as we know them today began to emerge as independent academic subjects during the early modern period, in the work of Newton and other 'natural philosophers', and numerous sub-disciplines developed during the centuries that followed. This part of the Cambridge Library Collection is devoted to landmark publications in this area which will be of interest to historians of science concerned with individual scientists, particular discoveries, and advances in scientific method, or with the establishment and development of scientific institutions around the world.

Studies in Spectrum Analysis

Sir Joseph Norman Lockyer (1836–1920) was one of the pioneers of astronomical spectroscopy and became one of the most influential astronomers of his time. His main interest was sun spectroscopy, which led him to discover helium independently of Pierre Janssen, a scientist who posited its existence in the same year. In addition to his work in astronomy, Lockyer was one of the founders of *Nature* and was the editor of the journal for its first fifty years. This is the second edition of Lockyer's guide to spectroscopy, first published in 1878. It begins with the basics of spectroscopy such as the physics of waves and the method of observing spectra. Later chapters describe the history of the method and some of Lockyer's own experiments and findings. This book is a fascinating part of the history of astronomy, giving insights into the development of a method vital to the field.

Cambridge University Press has long been a pioneer in the reissuing of out-of-print titles from its own backlist, producing digital reprints of books that are still sought after by scholars and students but could not be reprinted economically using traditional technology. The Cambridge Library Collection extends this activity to a wider range of books which are still of importance to researchers and professionals, either for the source material they contain, or as landmarks in the history of their academic discipline.

Drawing from the world-renowned collections in the Cambridge University Library, and guided by the advice of experts in each subject area, Cambridge University Press is using state-of-the-art scanning machines in its own Printing House to capture the content of each book selected for inclusion. The files are processed to give a consistently clear, crisp image, and the books finished to the high quality standard for which the Press is recognised around the world. The latest print-on-demand technology ensures that the books will remain available indefinitely, and that orders for single or multiple copies can quickly be supplied.

The Cambridge Library Collection will bring back to life books of enduring scholarly value (including out-of-copyright works originally issued by other publishers) across a wide range of disciplines in the humanities and social sciences and in science and technology.

THE

INTERNATIONAL SCIENTIFIC SERIES.

VOL. XXIII.

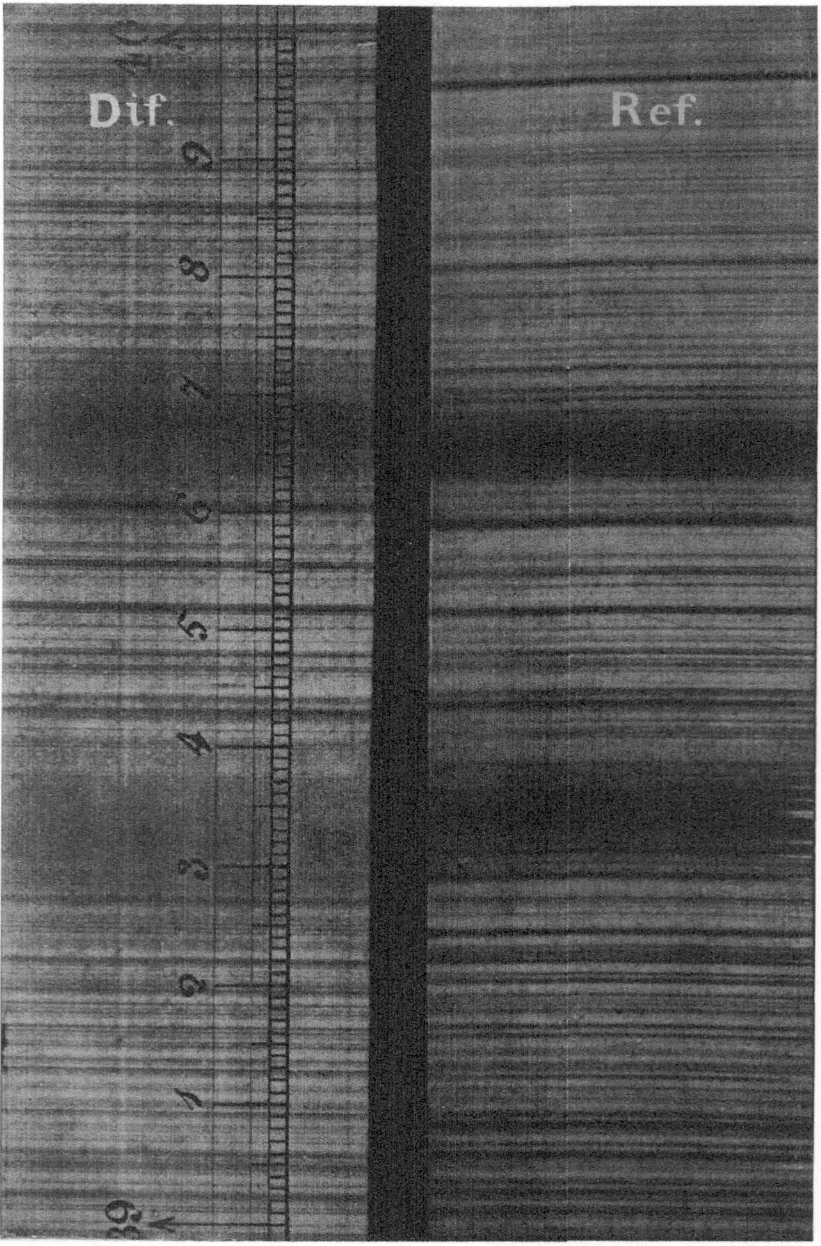
Dif.
Ref.

STUDIES IN

SPECTRUM ANALYSIS

BY

J. NORMAN LOCKYER, F.R.S.

CORRESPONDENT OF THE INSTITUTE OF FRANCE, ETC., ETC.

SECOND EDITION

LONDON

C. KEGAN PAUL & CO., 1, PATERNOSTER SQUARE

1878

CONTENTS.

———

CHAPTER I.

CHAPTER II.

CHAPTER III.

Contents. vii

LIST OF ILLUSTRATIONS.

N.B.—In the reproductions of the spectrum photographs the least refrangible end is in cases placed at top.

———— ++ ————

PLATES.

WOODCUTS.

———✦———

Woodcuts.   xi

STUDIES IN

SPECTRUM ANALYSIS.

———•———

CHAPTER I.

WAVES.

§ 1. *Preliminary.*

THE work of the true man of Science is a perpetual striving after a better and closer knowledge of the planet on which his lot is cast, and of the universe in the vastness of which that planet is lost.   The only way of doing this effectually, is to proceed as gradually, and therefore as surely as possible, along the dim untrodden ground lying beyond the known.   Such is scientific work.   There is no magic, no fetish in it.   There is no special class of men to whom it is given to become more familiar with the beauties and secrets of nature than another.   Each of us by his own work and thought, if he so choose, may enlarge the circle of his own knowledge at least, and thus make the universe more and more beautiful, to himself at all events, if not to others.

Futher, it now and then happens in the history of the human race upon this planet, that one particular generation gathers a rich harvest of this better and closer knowledge, this advancement generally coming from an exceeding small germ of thought.

Several such instances suggest themselves. How once a Dutchman experimenting with two spectacle-glasses produced the Telescope ; and how the field of the known and the knowable has been enlarged by the invention of that wonderful instrument. How once Sir Isaac Newton was in a garden and saw an apple fall ; and how the germ of thought which was started in his mind by that simple incident fructified into the theory of Universal Gravitation. Each step of this kind has more firmly knit the universe together, has welded it into a more and more perfect whole, and has enhanced the marvellous beauty of its structure.

Future times will say that either this, or perhaps the next, generation, is as favoured a one as that which saw the invention of the telescope or the immortal discovery of Newton : for as by the invention of the telescope the universe was almost infinitely extended ; as from Newton's discovery we learned that like energies were acting in like manner everywhere ; so in our time does the Spectroscope show us that like matter is acting in like manner everywhere ; so that if matter and energy be not identical, then these two, namely, matter and energy, may be termed the foundation stones of the universe in which we dwell.

The newer the science the more wary must be the

steps. In the newest of all the sciences, therefore, which enables us to regard matter wheresoever situate from an entirely new point of view, the investigator's caution must be redoubled, and those who would follow him must be careful to secure firm foothold at every step. Fortunately for us the laws and phenomena of nature have such a oneness in their diversity, and are so exquisitely intertwined, that it is possible for us in the consideration of any new branch of Science to aid our conceptions by mental images derived from the older sciences or ordinary phenomena, and this is especially true for that science now under consideration.

We can thus begin by some elementary notions which, when fully comprehended, will enable us to build on them conclusions which will be so many further steps.

By means of post-offices, railways, and electric telegraphs, we have the idea perpetually brought before us that in one place a man or a thing sends; that somewhere else, it may be near or it may be far off, we have a man or a thing which receives; and that between the man or the thing which sends, and the man or the thing which receives, there is a something which enables the thing sent to pass from one place to the other. There does not seem to be any deep science in this, nor is there; but these considerations enable us to make an important distinction. In the case of two boys playing at ball, one boy throwing the ball to the other, we have also a sender and a receiver, and the thing sent goes bodily from the one who sends to the one who receives.

So in a parcel sent by train, but not so in the case of a telegraphic message. In the electric telegraph office two instruments may be seen—one the *receiving* instrument, the other the *sender.* Between the office in which we may be and the office with which communication is being made, there is a wire. We know that a thing is not sent bodily along that wire in the same way as the boy sends the ball to his fellow, or as the goods train carries the parcel. We have there in fact a condition of motion with which science at present is not absolutely familiar ; but we picture what happens by supposing that we have *a state of things* which travels. The wire must be there to carry the message, and yet the wire does not carry the message in the same way as a train carries a parcel.

Take another case. I burn my foot, I instantly raise it. To make me conscious that my foot had been burnt, a message (as we know now) must have gone from my foot to my brain, and a return message must have gone from my brain to my foot, to tell it to change its position so as not to be burnt any more. Now it is generally held that this internal transit of messages is not managed by electricity, but that although electricity is not here at work, still that there is something which behaves very much after the manner of electricity. No one imagines that the *pain* travels up the leg and then back again ; it is, in fact, a *state of things* which travels up from the nerve of the foot to the brain ; and then there is another *state of things* which travels back again from the brain to the foot, along another set of nerves. A rope will here afford us a useful mental image. By shaking a rope we can send that *state of things* we call a wave along it without the rope itself travelling as a whole; this will help to give us an idea of what is meant when we say that a state of things travels along a wire or along a nerve, and brings about either those electrical disturbances which result in the conveyance of a message, or that nerve action which generates the action of the brain.

§ 2. *Water Waves.*

Next to dwell more especially upon the word *wave*, and the idea which that word most generally calls forth. Let us find a piece of tranquil water and drop a stone into it. What happens ?—a most beautiful thing, full of the most precious teachings. The place where the stone fell in is immediately surrounded by what we all recognize as a wave of water travelling outwards, and then another is generated, and then another, until at length an exquisite series of concentric waves is 'seen, all apparently travelling outwards—not with uncertain speed, but so regularly that all the waves all round are all parts of circles and of concentric circles.

Let us drop two stones in at some little distance apart. What happens then ? We have two similar systems each working its way outwards, to all appearance independently of the other. We get what is represented in Fig. 1.

Now these appearances are as if there were an actual outpouring of water from the cavity made by the stone ; but if we strew small pieces of paper or other light material on the water surface before we drop the stone, we find that it is not the water which moves outwards,

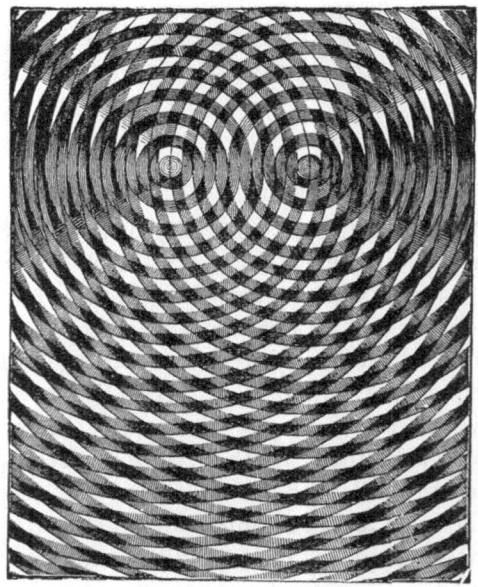

Fig. 1.—Superposition of two wave-systems.

but only the state of things—the wave. Each particle of water moves in a circular or elliptic path in a vertical plane lying along the direction of the wave, and so comes again to its original place.   Hence it is that only the *phase* goes on—*how* it goes on will easily be gathered from Fig. 2.

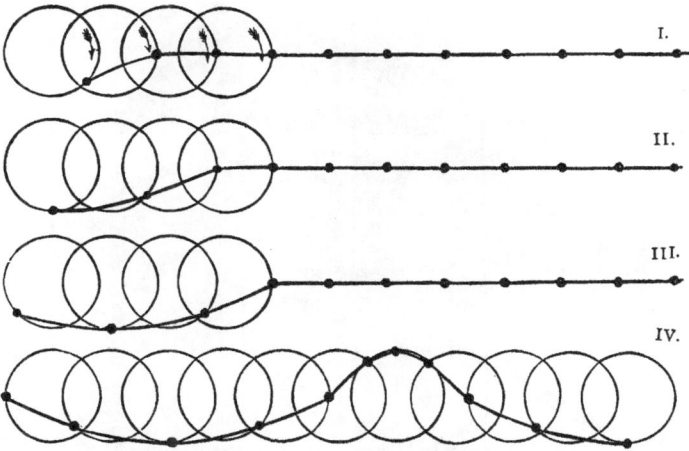

FIG. 2.—Showing the formation of waves by the circular motion of each particle of water in a vertical plane. Eight positions in each revolution are shown.

I. One particle in motion.—II. Two particles in motion.—III. Three particles in motion.—IV. Complete wave and motion of its elements.

§ 3. *Sound Waves.*

Let us now pass to a disturbance of another kind, from two dimensions to three, from the surface of water to air.

We hear the report of a gun or the screech of a railway whistle, or any other noise which strikes the ear. How comes it that the ear is struck? Certainly no one will imagine that the sound comes from the

cannon or from the railway whistle like a mighty rush
of air.   If it came like a wind we should feel it as

3.—Graphical method of observing the mode of vibration of a tuning fork.

a wind, but as a matter of fact no rush of this kind is
felt.   It is clear, therefore, that we do not get a bodily

transmission, so to speak, as we get it in the case of the ball thrown from one boy to the other. We have a *state of things* passing from the sender of the sound

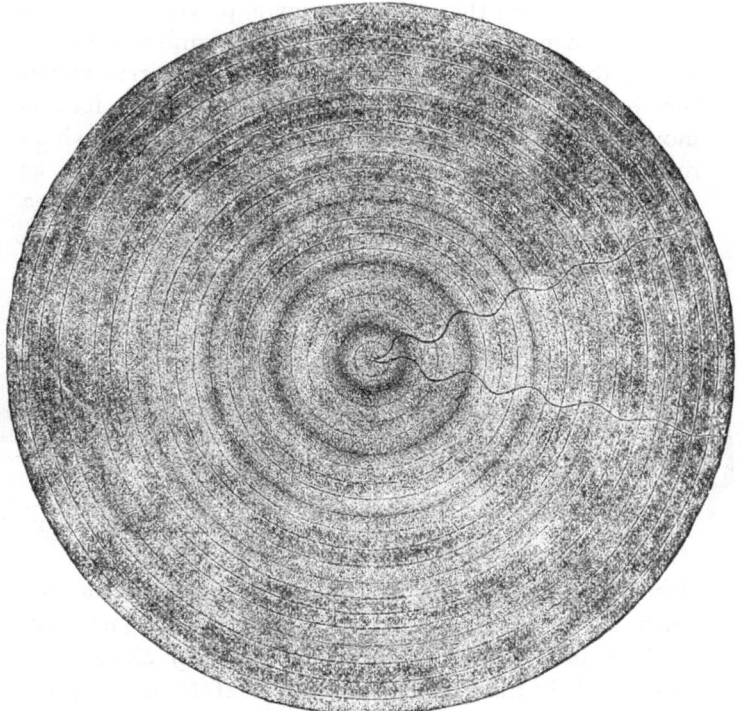

FIG. 4.—Shells of compressed and rarefied air produced by a source of sound.

to the receiver ; the medium through which the sound passes being the air. A sounding body in the middle of a room for instance, must send out shells of sound,

as it were, in all directions, because people above, below,
and all round it would hear the sound. Replace the
stone by a tuning fork. To one prong of this fasten
a mirror, and on this mirror throw a powerful beam of
light. When this tuning fork is bowed, and a sound is
heard, the light thrown by the attached mirror shows
the fork to be vibrating, and when the tuning fork is
moved we get an appearance on the screen which
reminds us of the rope, or we may use the fork as
shown in Fig. 3, and obtain a wavy record on a
blackened cylinder.

Experiment shows that we have at one time a
sphere of compression—that is to say the air is packed

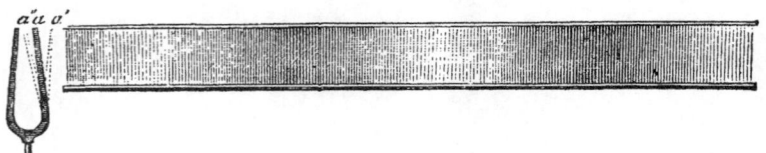

Fig. 5.—Propagation of sound waves along a cylinder.

closely together; and, again, a sphere of rarefaction,
when the particles of air are torn further apart than
they are in the other position. The *state of things* then
that travels in the case of sound, is a state of compression and rarefaction of the air. Hence the particle of
air travels differently from the particle of water; it
moves backwards and forwards in a straight line in the
direction in which the sound is propagated.

The annexed figure will show how this backward-

and-forward movement results in the compressions and rarefactions to which reference has been made, in consequence of the impulse having been imparted to one molecule after the other. Owing to the pendulum-like motion of the molecules, their relative positions vary at each instant of time.

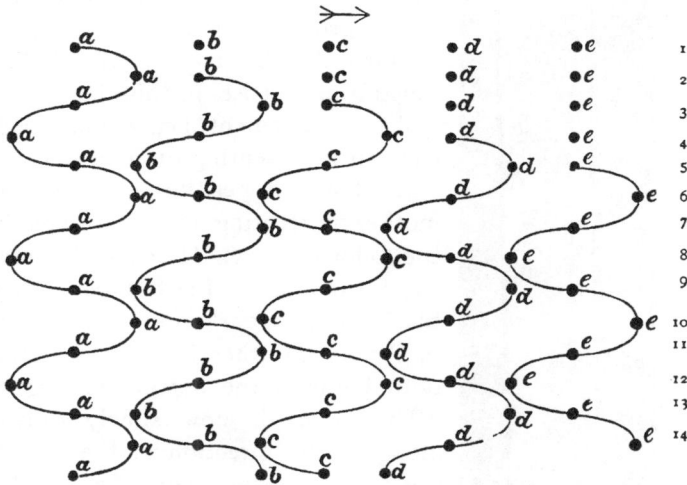

FIG. 6.—Sound Waves. Particles of air, *a*, *b*, *c*, *d*, *e*, are in position 1 at rest. The remaining positions show how they are situated at successive instants, when a continuous series of impulses reaches them from the left. In position 2, *e.g.*, only one particle has begun its oscillation. In position 3, only two, while in position 6, all are in motion.

Professor Weinhold has given in his "Experimental Physics" a good method of obtaining on a plane a mental image of what goes on in a so-called sound wave, and by the courtesy of Messrs. Longmans I am enabled to give here the illustrations which he employs.

After all the particles have been put into motion as shown in Fig. 6, if we graphically represent the backward and forward oscillation of a particle by such a wavy line as in Fig. 7, we shall, when we put a large

number of such waves side by side introducing the change of phase, have such an arrangement of wavy lines as is represented in Fig. 8.

Now the beauty of Weinhold's illustration consists in this; he almost makes each element of each line—each element representing of course a particle of air—appear to be actually in motion by treating the above figs. in the following way. He cuts a narrow slit *SS* in a piece of stiff paper, either black or of a dark colour as shown in Fig. 9. He then holds this on the dotted line at the bottom of Fig. 7. "The book is now slowly drawn along in the direction of the arrow, the piece of paper being held in the same position. At first the lower

FIG. 7.

extremity of the curved line in *A* is seen through the slit; but as the book is drawn along, the portions to the right and those to the left come successively in view; the small white dot which is the only visible portion of the curved line, appears as a point which moves first to the right and then to the left, and imitates closely the motion of a vibrating particle of air, the rate of motion being, however, much slower. If now the slit be placed over the dotted line [at the

bottom of Fig. 8] and the book drawn along underneath it in the direction of the arrow, a representation is obtained of the motion of a series of particles of air which are acted on by a number of successive equal undulations or waves. Each particle merely moves a little right and left and always comes back again to its starting point; but the condensations and rarefactions, represented by the lines being respectively closer together or further apart, are gradually transmitted through the whole series of air particles from one end to the other." *

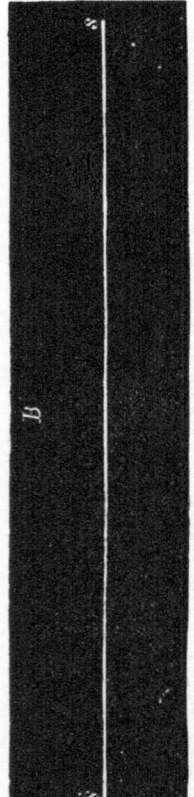

Fig. 9.

In dwelling upon sound phenomena, we have the advantage of dealing with phenomena about which science says she does know something: from a consideration of these known facts we shall be able, slowly, but surely, to grasp some of the much less familiar phenomena with which spectrum analysis is especially concerned.

We all know that some sounds are what is termed high and others low, a difference which in scientific language is expressed by saying that sounds have a difference in pitch. We know that the difference between a sound which is pitched high and a sound which is pitched low

* "Experimental Physics," p. 332.

is simply this, that the pulses or waves, as we may call
them for simplicity's sake, which go from the sender-
forth of the sound (which may be a cannon, a piano, or
anything else) to the receiver, which is generally the
human ear, are of different lengths. What in physics is
called a sound wave is constructed as follows : We have
a line AX, which represents the normal condition of the
air through which the sound is travelling, and curves

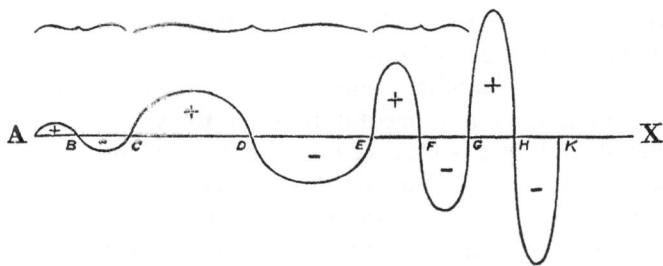

Fig. 10.—Sound waves of different lengths and amplitudes.

which represent to the eye—first, the relative amounts
of compression (+) and rarefaction (—) brought about
by the sound in the case of each pulse, and secondly the
relationship of this to the actual length of the wave, or,
what is the same thing, the time taken for the pulse to
travel. Thus we may have long waves and short waves
independently of the amount of compression or rarefaction, and much or little compression or rarefaction independently of the length of the wave. We know that the
difference between a high note and a low note, whether

of the voice or of a musical instrument, is, that the high note we can prove to be produced by a succession of *short* waves—such pulses as have been described— and the low note by a succession of long waves.

Now the loudness or softness of a note does not alter its pitch, that is, it does not alter the length of its waves or the rate at which they travel. I can send a wave along the rope either violently or gently, but with the same tension of the rope we shall find that the length of the waves is the same, provided the period of vibration is the same. Hence then the other idea added to the idea of pitch.

There is another point which is worth noting, although it is not needful to refer to it in any great detail, and that is that we know that sound travels with a certain velocity, and that this rate is subject to certain small variations owing to different causes.

We not only have to deal with amplitude—that is the departure of the + and − parts of the curve from the line AX—and velocity, but we have this most impor- tant and very beautiful fact (for fact it is), which some will have observed for themselves. If a person in a room in which there is a piano sings a note, the string of a piano tuned to that particular note will respond, and if he sing another note, then another string will reply, the first string being silent. And if the experi- menter were skilled enough to sing one by one all the notes to which the strings of the piano are tuned, all the strings would be set into vibration one by one, note for note. This fact may be explained in this way ; a piano wire, or similar sonorous body, which is constructed to do a certain thing—in this case to sound a particular note—always sounds that note when it is called upon *in a proper way* to do it. Now this is the point. The proper way may be either (1) that a particular vibration should fall upon it, or (2) that it should be set to work to generate that vibration in itself. If the piano wire only gives the same sound when struck either hard or soft, it is because it is manufactured to do one particular kind of work, and it can do no other.

Now we may pass from a piano wire to a tuning fork. We find that by using different quantities, or different shapes, of metal, these instruments give out different notes. If all be of the same metal, the different quantities of metal will give us a difference in the pitch. This demonstrates that the pitch of a note is independent of any particular quality of the substance set into vibration. Now although a great many musical instruments can sound the same note, yet the music, the *tone*, which one gets out of them is very different. That is, the pitch being the same, the quality of the note changes because the wave, or rather the system of waves, which we obtain is different. For instance, if we sound a note upon the violin, or the French horn, or the flute, or the clarionet, anybody who knows anything of music will tell which is in question, so that here we have in addition to wave length and wave amplitude another attribute, namely, that which in French is called "timbre," in German "Klangfarbe," and in English, "tone" or "quality."

C

To sum up then what we have already stated with regard to sound. When we deal with the phenomena of sound, we find that they are composed of disturbances or vibrations connecting the sender with the receiver ; that the sound may vary in pitch ; that the amount of the sound depends upon the amplitude ; that the sound is independent of the material of the sender or the kind of disturbance, so far as pitch goes, but that so far as *quality* is concerned it is to a certain extent dependent upon the nature of the material and of the kind of disturbance.

§ 4. *Light Waves.*

We have now to consider that kind of disturbance to which we owe the sensation of light—light being to the eyes of the human race very much what sound is to the ears. Here again, for simplicity's sake, let us look at the question in this three-fold point of view. Let us deal with the sender, the receiver, and the medium which connects the sender with the receiver ; first observing that so far as we know at present, not to go too much into detail, there are three kinds of receivers. There is first of all that marvellous instrument the human eye. There is next, also a very marvellous thing, the photographic plate.

How is it that a few words will awaken in each one of us many memories of our childhood ? Because we saw certain things in our childhood, by means of our eyes; and the impressions which we received by means of our eyes were recorded in our brains, and we possess the faculty of being able to build them up— to recollect them—again. We have there a permanent method, so to speak, of recording things which are seen by the eye —of recording messages from a certain sort of sender. In the photographic plate, we have also a permanent record of a certain condition of things, whether a face, a house, or a ship, or a particular state of the sea or sky; presented to a particular set of chemical conditions at some past time, which brings back some pleasant remembrance of friends perhaps far away. There we have two receivers which more or less accurately, and more or less permanently, record the disturbance which once impinged upon them.

Then besides the eye and the photographic plate we have everything else in nature—the houses we live in, the furniture, the familiar faces around us, this book and everything else on the planet—and not only these but every thing in the Cosmos which does not shine by its own light; these form the third class of receivers, that is to say, those which do not record, at all events obviously, the impressions made upon them, and which, more or less perfectly, reflect light.

So much therefore for receivers of this kind of vibration. We must bear in mind that at night, or in a dark room the things mentioned and such like become invisible. Our eyes fail to see them, a fact which shows that the receiver plays a very important part. On the other hand everything in a bright summer's day receives light from

one light source—the sun.   How is it then that with the
first class of receiver—the eye—we are enabled, unless
indeed we be colour blind, to see all the beautiful and
glorious varieties of nature in its ten thousandfold hues ;
while the other receiver, the photographic plate, gives us
but black and white?   Why are roses red, and why are
leaves green ?   There is the same light in the sky, and
the same absence of form—the same absence of visibility—in the dark ; yet with the light coming from one
and the same light-source we get all these different
effects.   How is this ?   It drives us to the conclusion,
either that the receivers, to which our attention has
been particularly directed, deal with light in very
different ways, or that by some means or other they
manage to get hold of different kinds of light.

Here then we must seek for some explanation of the
various colours that we see in nature.   We have referred
to the receivers, including those that reflect the light
which they receive ; now let us consider the things
which send out the light.   Among these are the sun, the
moon, the stars, gas, and candles which are most familiar
to us as sending out light.   And it will be well to remark
here, and the reason why will be clear by-and-by, that
the light which all these senders give to us is white
light in the main.   But we get other kinds of light.

We have, for instance, that of the electric lamp—a
very powerful source of light, only a very little less
powerful (as some people think) than the sun itself.
It proceeds from two carbon poles which are rendered
intensely incandescent by means of an electric current.

By inserting different metals between these poles, we find that we get light not only from the poles of the lamp itself (a source of white light), but that we obtain various coloured phenomena by this addition.

It is not alone by means of the electric arc, or spark, that these phenomena can be produced ; on putting salts of different metals into the flame of the Bunsen burner, we observe that the colour of the flame will depend upon the substances put into it. Sodium will give us a yellow flame, lithium will impart a certain redness to the flame, and thallium a green tinge. Now, if instead of dealing with metallic salts, we prefer to take certain gases, and render them brightly luminous or glowing, by means of the passage of an electric current, we shall in that case also get differently coloured effects. Some of these gases are red, some are green, some are violet, and so on.

Now these coloured phenomena of which we have spoken, are things which we can and do produce with chemical or physical instruments ; but in addition to those, we have various colour-giving bodies in the skies, in the same way as we have the sun, the moon and those stars which are not brilliantly coloured. All who were fortunate enough to see that beautiful comet which was visible in July 1874, must have noticed that it was a yellow-looking comet—not so yellow as the sodium flame, but still distinctly yellow. Those who have had the opportunity of observing some of the stars through a telescope, or, what is nearly as good, those who have been across the Line and seen some of the stars of the Southern Hemisphere, know that some of the stars in the heavens are as beautiful, and, so to speak, as majestic for their colour, as others are for their brilliancy. Again, those who have seen a total solar eclipse will have seen a large and interesting portion of the sun which we cannot see at any other time—a region of beautiful colours as well as of grotesque forms.

So that both in the heavens and on the earth we get instances of light which is white, and of light which is coloured.

So much for the senders. Now one word about the medium ; for, as we shall understand, in the case of light as in the case of electricity about which we are uncertain, and as in the case of sound about which we are absolutely certain, there is no transmission of anything but a state or a condition of things, a disturbance or a vibration, between the sender and the receiver. The light, for instance, which appears to be given out by a candle, and which is received by our eyes, does not come bodily from that candle, like so many small bullets. The sender — in this case the candle — is simply a something which puts something else into motion. And then there is a something which conveys that motion. By striking a bell and ringing it, a noise may be made ; but if that bell is put into a glass vessel, and the air exhausted, and the bell is then rung, we would not hear it at all. How is this ? Because the carrier of the sound waves is the air; and when we take the air away we take away all chance of getting sound transmitted from one place to another. We know, for instance, that in our moon there is absolutely no sound. If the moon were teeming with men to-morrow no one could hear another person speak. No sound, either loud or soft, could be heard by any inhabitant of the moon, because the moon practically has no atmosphere.

Still, notwithstanding that there is no air all the way between us and the moon, or all the way between us and the sun, yet we get light from the moon and from the sun. How, then, is this?

§ 5. *The Theory of Ether.*

Physicists imagine that there is a something which they call "ether," infinitely less gross in structure than air, which pervades all nature and permeates all bodies ; and that the disturbance or light wave produced by a light sender, or radiator, is transmitted along the ether very much in the same way as the wave "state" is transmitted along water, or the state of motion is transmitted along a rope. Associated with this ether we have the undulatory theory of light, which supposes that everything which sends out light sets the ether— this subtle, imponderable air, so to speak—in vibration ; and that those vibrations travel without any transmission of the substance of the ether from each sender of light to each receiver of light. Here we have one of the great triumphs of modern science, because, as many of us know, so great a man as Sir Isaac Newton held (and he was quite justified in so doing.

with the facts at his disposal in his day), what was called the "corpuscular" theory of light, which supposed that little shots of light, so to speak, like little shots out of a cannon, were emitted from every sender out of light; in fact, that the ether carried light as a train carries a parcel, and not as a telegraphic wire carries a message. That, however, is not the opinion which men of science hold now. They have changed that opinion because their basis of facts has been enlarged. Such must ever be the conditions of science, and science can never be so flourishing as when she is changing her opinions, because her opinions can never be changed unless she has acquired a number of new truths.

Although, then, it is not generally supposed that there is anything in the nature of an atmosphere extending all the way between us and the sun, yet, because we see the sun, we suppose that there is some medium present, which medium has been named the ether. As there are ninety-one millions of miles, or so, between us and the sun, and ninety-one millions of miles multiplied millions of times, between us and some of the stars that we can see, we are bound to concede that this medium is almost, if not quite, perfect in its capacity for transmitting light, and does not make the light pay any appreciable toll on its passage. We know that the atmosphere is sometimes so heterogeneous as to density that sound travels along it with very great difficulty. This idea will enable you to appreciate the other—that light can have no great difficulty in travelling across the ether, seeing that it reaches us from stars immensely distant. We may, therefore, say that in the case of light, we have ether as a general and almost perfect medium or transmitter of the disturbance produced by a radiating body to those various classes of receivers to which attention has already been drawn.

How, then, are we to picture to ourselves the motions of the particles of ether in a light wave? We are already familiar with the circular orbits of the molecules of a water wave in a vertical plane in the direction of motion, and of the forward and backward motion of a particle of air in the direction of motion of a sound disturbance. The motion of the particles of ether, as imagined by modern physicists, is widely different.

In the first place the motion is transverse to the path of the disturbance—that is, the vibrations take place in planes perpendicular to the direction of the ray.

What, then, is the motion of the etherial molecules in this plane? It varies, depending doubtless upon the vibration of the sender. The molecule may describe a straight path or an orbit—*i.e.*, its path may be straight, circular, or elliptical—but in all cases the path or orbit lies in a plane at right angles to the direction of the ray.

A row of balls in a straight line may be taken to represent particles of ether at rest. If we imagine the balls to start successively and vibrate uniformly up and down we shall get a wave system finally established along the whole line ; we shall have crests and hollows, and we at once get the same introduction of the ideas of wave length (the length from hollow to hollow or crest to crest), and of amplitude, as we got in the case of the sound waves.

Here, then, we have one form in which the mutual

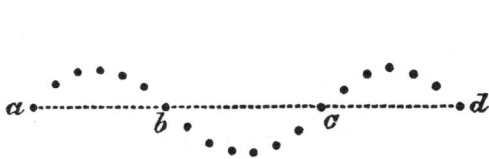

FIG. 11.—Wave Systems.

attraction or elastic cohesion of the etherial particles conveys a disturbance.

Now in ordinary light the paths and orbits are not all similarly situated. That is, the straight lines described by the particles represented in Fig 11 may pass through the central line *a d* at different angles, and the major axes of the orbits of those which have elliptic paths may also cut the central line at different angles ; so that, to quote Mr. Spottiswoode,* "although there is reason to believe that in general the orbits of a considerable number of consecutive molecules may be similarly situated, yet in a finite portion of the ray there is a sufficient number of variations of situation to prevent any preponderance of average direction."

* " Polarization of Light," p. 6.

§ 6. *Wave Lengths.*

A word now as to the length of light waves, so that the scale on which the motions of the molecules of ether —our medium—take place may be comprehended. A comparison with the waves of sound will again bring out other similarities between the two classes of phenomena.

First, then, with regard to sound. The average velocity with which a sound disturbance is propagated through the air is 1,140 feet in each second. It has been demonstrated by experiment that the lowest effective note we can appreciate as music is one in which the disturbances enter the ear at the rate of 16½ per second.

Imagine then a column of air 1,140 feet long with sixteen compressions and rarefactions along the length. It is clear that this whole wave system must beat upon our ears each second, and that the length of the wave, *i.e.*, the distance from maximum compression to maximum compression, or from minimum rarefaction to minimum rarefaction must be nearly 70 feet.

The highest appreciable note according to Helmholtz is one with 38,000 vibrations per second.

Between these extreme limits, then, we have all the glorious world of musical sound which our ears are tuned to appreciate. The air is possibly teeming with sounds both below and above our range.

Now as regards light waves. As the ether is infinitely more elastic than our grosser air, so are the disturbances propagated with a velocity which quite baffles our comprehension. The latest measure ments tell us that a light-producing disturbance travels at the rate of 186,000 miles in a second of time. Imagine the molecular agitation depending upon this statement, and then remember that a glowworm can set it all going, and that, when once in full swing, the distance of the most remote star is traversed as it were at a bound and without sensible loss of energy.

Then as to the dimensions of the light disturbance. The length of the longest wave that we can see is 00076009 of a millimetre * (76,009 hundred-millionths of a millimetre or about $\frac{1}{39000}$ of an inch). The length of the shortest is 00039328 of a millimetre (39,328 hundred-millionths of a millimetre or about $\frac{1}{37000}$ of an inch). The longest visible waves are red, the shortest violet.

Now as in 186,000 miles there are 298,000,000 metres, or 29,800,000,000,000,000,000 hundred-millionths of a millimetre, and as all the waves must enter the eye in a second, we have for the number of wave crests per second

$$\frac{29,800,000,000,000,000,000}{76009} = 392,000,000,000,000$$

* A millimetre is 0·03937 of an inch.

that is 392 billions of waves entering our eye each second in the case of red light, and

$$\frac{29,800,000,000,000,000,000}{39,328} = 757,000,000,000,000$$

that is 757 billions in the case of violet light.

§ 7. *Interference of Matter with the Motions of the Ether.*

We must next observe that light is not necessarily limited to transmission through the ether in free space. If a glass of port wine is held up to the sun the light passes through it and seems red. In that case the light has had to pass through the ether *plus* the port wine, and there we can see that the new medium has made an enormous difference in the light which was originally sent us.

Supposing the light from an electric lamp were thrown upon a screen, we should see that it is white, that is, the same kind of light as we obtain from the sun. Imagine that the light is really coming from the sun; by interposing a piece of blue or red glass (adding these substances to the ether, as it were), we at once alter the condition of things, and get a blue or a red light upon the screen. So it is clear, that if we want to study light phenomena completely we must not only take into account the different circumstances connected with the sender and with the receiver, but also the different circumstances connected with the media through which the light passes, or, as we shall see by and by, with those media which *absorb* light; for, although we do not know that ether absorbs the light, yet practically we know that everything else does. We know the redness of the sun at evening arises, not from absorption by the ether, but from absorption by a great thickness of our atmosphere, which practically does for the light of the sun what the piece of red glass did for the light of the electric arc in the experiment above suggested.

We see then, still dealing with our complicated medium (that is ether + matter in some cases), that their association leads to an absorption of light so that the receiver does not get all the disturbance set up by the sender, in consequence of the vibrations of the ether being affected by the molecules of the various bodies through which they have to pass.

This result is not the only one which follows from the entanglement, so to speak, of the ether waves among the molecules of matter. If the disturbance is travelling in such a direction that it passes into a substance denser than air—such as water or glass—otherwise than perpendicularly to the surface, the direction of propagation of disturbance is changed, the wave, so to speak, has changed front, and the greater the difference between the density of the two kinds of matter, such as air and water, or air and diamond, thus passed through, the greater will be this change of front, and the more will the direction in which the light travels be changed.

§ 8. *Action of the Prism.*

But the change of front is accompanied by something else which is very much more important for our present purpose, and this can be studied best when we make the disturbance enter and leave the surfaces of the body built up of the denser molecules at the same angle.

This can be accomplished by using glass as an

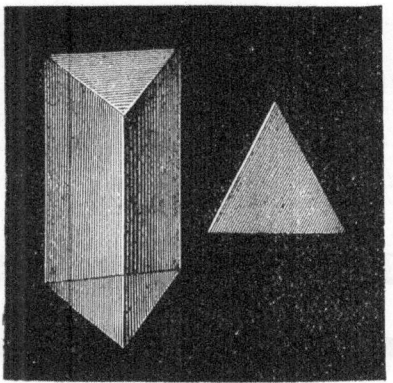

Fig. 12.—The prism.

illustration of the material addition to the medium, and shaping it into the form of a prism.

The effect which is produced was first described by Kepler, and explained by Newton ; but it has required the undulatory theory of light to render a complete understanding of it possible.

The addition of the molecules of glass, presented, in the way indicated, to the ether disturbance, has resulted (1) in turning the ray out of its course, and if it be a ray of white light (2) in splitting it up into its constituents, each constituent being represented by a different colour, or (3) if the ray be of any special pure colour, in causing it to travel in a direction which is constant for the same colour, but different for different colours.

Glass affords us an instance in which the dispersion of colour thus obtained is *normal*, that is, in the order of wave lengths. The order of the colours obtained is as follows :—

Red, orange, yellow, green, blue, violet, indigo.

But there are substances the action of the molecules of which upon the ether is very different, and we get *abnormal dispersion*, so called because the above order is changed.

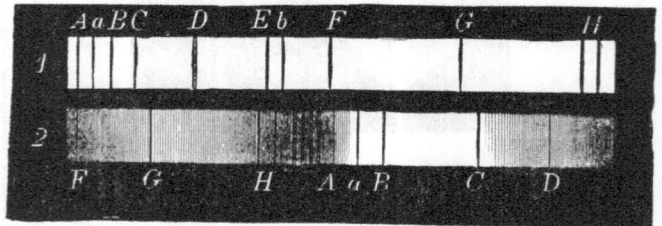

Fig. 13.—(1) Normal dispersion of glass.—(2) Abnormal dispersion of Fuchsine.— *A a B C*, Red.—*D*, Yellow.—*E F*, Green.—*G*, Blue.—*H*, Violet.

The prism tells us that a beam of white light is not a simple thing, but that it may be likened to a rope with an infinite number of strands. If, for instance, by some concerted action all the keys of a piano are pressed down, a certain sound results, made up of a combination of all the sounds upon the keyboard. This then is the sound analogous to a ray of white light. The reasoning which lies at the bottom of all the new researches which have made us as familiar with matter millions and millions of miles from us as we are with the matter around us, arises from the perfect establishment of the idea, that a ray of white light is universally composed of waves of light of various lengths, just as that noise from the piano was also composed of different true musical notes, that is to say of waves of sound of various lengths ; and that each light of special colour is composed of a single wave length or of a special combination of wave lengths.

If then instead of letting the white light which we get from the sun or the electric lamp, travel through a fine slit straight from the sender to the receiver, we insert a prism and lens in its path, we observe an effect of a complex nature; the light is thrown out of its course, and instead of the lens forming a single image of the slit through which it emerged, as it did before— instead of the image of the slit, which was white and small before—we shall have a rainbow coloured image stretching across the screen. By adding a second prism to aid the action of the first, we get the same effect increased, as might be expected. That rainbow coloured band is the spectrum.

Now, the difference between the blue light at one end of the beautifully coloured band and the red at the other, is nothing more nor less than a difference almost identical with the difference between a high note and a low note upon the piano. The reason why one end of the coloured band, which in what follows will be called the spectrum, is red, and the other blue, is that in light as in sound we have a system of disturbances or waves ; we have long waves and short waves, and what the low notes are to music the red waves are to light, and what the high notes are to music the blue waves are to light.

There is a strict analogy between the world of sound and the world of light. Ears are tuned to hear different sounds—some people can hear much higher notes than others, and some people can hear much lower notes than others. In the same way some people can see colours to which other people are blind; indeed, the more we go into this matter, and the more complete we make our enquiries, the more striking becomes the analogy between these two classes of phenomena.

Hence it is that I have attempted to utilize the phenomena of sound as a sort of subsoil plough, to enable us better to cultivate those fields which modern science has annexed to the region of the known—fields wonderfully rich in facts dealing not only with the qualities of matter, but with the physical bases of matter itself; with this beautiful and undreamt of expansion, that it is indifferent whether that matter is in the hand of the experimenter in his own laboratory, or whether it is sending out light to us upon this earth from the very confines of the universe. Nature is so absolutely and universally true and regular in all that she does, and modern science is of itself such a slight regarder of time and space, that when it is a question of studying the smaller aggregations of matter the spectroscope enables us to tell not only what kind of matter is at work, but it tells us a great deal, and will tell us a great deal more, about the actual conditions of that matter. Indeed, it is probable that in a few years we may know very much more about matter very far removed from our own planet, than we do of a great deal of it on the very planet itself on which we dwell.

§ 9. *Recapitulation.*

Let us assume that we are now prepared to take what we know about sound as representing, with more or less accuracy, some of the things that we know about light, and recapitulate the points which have already been touched on. Both with regard to sound and light we may consider different substances, first as senders, then as receivers, and then as media. First, as to the senders with regard to sound—sound is set up or produced by bodies such as a tuning fork, and we know that sound is due to the vibrations or oscillations of that tuning fork imparting a regular disturbance to the air ; the sound which that or any other body produces depending upon the kind of disturbance which it sets up.

Now with regard to light sources, a body which gives out light, does for light exactly what the tuning fork does for sound.   A bell ringing is the equivalent of a fire burning or a star shining.   Both with regard to sound and to light there are various kinds of receivers. We can, for instance, by preparing certain surfaces, receive and place on record the shape and length of waves of sound—we can make a sound disturbance permanent.   Photography provides a means of securing a permanent record of light-disturbances.   Here we have two receivers, one of sound, the other of light, which give a more or less permanent record.

Now I need not say, that as men and women we have ears and eyes, the ear doing for the wave of sound what the eye does for the wave of light.

With regard to the medium—always to keep to our phraseology—we have the air, whose function it is to transmit waves of sound : and we have the ether to transmit to us the waves of light.

We can imagine a compound sound composed of notes of all possible pitch ; we have an exact equivalent of this in the case of the light in the continuous spectrum, one in which from the red end to the blue or violet end there is no break in the light ; like an army going into action there are no vacant places in the line.

If we press down first one note of the piano and then another, we get an effect due not to a complete mixture of all possible sounds, but to each sound by itself.   Now spectrum analysis depends upon this, that what any one note of a piano which you choose to touch does for sound, each particle of matter does for light.

Experiment has shown us that the "light-note" so to speak, given out by the simplest particles of different kinds of matter, differs for each kind of matter. If we examine the spectrum of the light sent out by particles in a state of vapour, such as the vapour of sodium for example, we shall have the equivalent of what we get when we strike a single note upon the piano. We have a spectrum composed principally of a very decided line in the yellow. It is very important that the connection between the yellow line and the single note of the piano, and between the continuous spectrum and the sound produced by sounding all the notes of the piano at once, should be perfectly understood. Suppose we now take a metal which gives us a line not in the yellow but in the green ; the metal thallium. What, it may be asked, is the difference between the light being in the yellow and the light being in the green? The quality of the "light note" of thallium is different, so to speak, from the quality

of the light note of sodium, as 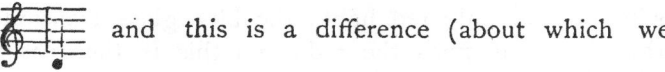 is different from

and this is a difference (about which we know very little) which enables us to tell in a moment whether we have to do with sodium or thallium, when we make each vapour send out its light.

We have already got out two very different characteristics among our light senders. We have first of all, that light source which gives us a continuous spectrum, that is a series of waves quite complete so far as the simple spectrum goes, and we have next that particular kind of light source which, instead of giving us a continuous spectrum, affords us one with bright lines, that is to say, parts only of the complete spectrum are represented in the light, because parts only of a complete system of waves are given out.

§ 10. *Absorption of Light.*

We have already seen that the medium which a light disturbance employs to get to us is the ether, and the ether has no effect upon light except to transmit it; that if in the path of the light which is sent to us, and received by us, we place something else besides the ether, then we may to a very large extent modify the qualities, so to speak, of the original disturbance.

By superadding the transmission through glass coloured red or blue, to the transmission through the ether, we get a distinct difference in the effect. In the red glass something is introduced in addition to the ether, which will transmit only red light; the blue glass transmits the blue and stops the red—and this is the reason why blue glass appears blue.

Here we are dealing with a class of experiments which provide us with what are termed absorption phenomena; that is, the differences are due not to the sender but to the medium, and the medium never adds, it always subtracts or, as it is termed, absorbs. If instead of using coloured glasses, we take a solution of

potassic permanganate—we shall observe certain dark
bars across the spectrum, indicating that there is in
nature a class of bodies which have this very distinct
effect upon the spectrum. Another experiment
will enable us to get a much more definite effect. It
will be recollected that sodium vapour was the vapour
which, when added to the flame of the Bunsen burner,
gave an intensely yellow light. Let us study the effect
of using sodium vapour as the medium—not as a source
of light but as an absorber. This we can do by sending
the white light of the electric arc through some sodium
vapour as well as the prism upon its way to the screen.
In place of the bright yellow line we saw before, we
shall see a dark line upon the screen.

This experiment gives us an idea of a class of spectra
of which we have very few natural representatives upon
this earth, in consequence most probably of the complicated molecular conditions found in a cool planet—a
class for which we have to search the skies, and which
we can find in almost every star which shines on the
face of heaven.

Here again an analogy drawn from sound will help us.

Suppose we have a long room and a fiddle at one
end of it, and that between it and an observer at the
other end of the room there is a screen of fiddles,
all tuned like the solitary one. We know that in that
case, as a matter of fact, the observer would scarcely
hear the note produced upon any one of the open strings
of that fiddle. Why? The reason is that the open
strings of this fiddle, in unison with all the other fiddles,

would set all the other open strings corresponding to it also vibrating, and upon the principle that you cannot eat your cake and have it too, the vibration of the fiddle cannot set all those strings vibrating, and still pass on to the other end of the room as if nothing had happened.

The work, in fact, which the air, the medium in this case, would have to do to make its vibration audible to the observer, would be locally done, so to speak, upon the screen of fiddles ; the work done would decrease the amplitude of the vibration, and the sound would be weakened.

Now this, as Professors Stokes and Ångström were the first to point out, is the real explanation of the result above mentioned.

Here we have a striking parallel instance of the fact that light phenomena are due to vibrations of light sources, communicated to us not by anything coming bodily from the light source, but by corresponding vibrations set up in the mysterious ether. If a sound wave travelling along the air, or a light wave travelling along the ether, finds in its path a vibrating body which is ready to receive the vibration, *whether it be already vibrating sufficiently to give us the impression of light or not*, that vibration is arrested or lessened, the sympathising body taking up the vibration in whole or in part.

§ 11. *The Basis of Spectrum Analysis.*

Light senders are really particles of bodies in vibration, and if there be no vibration there will be no sending out of light. The reason why things such as gas, flames, candles, the sun, and other bodies send us out light is this, that they are in a state of energetic vibration—in that state which we generally call hot.

The hotter a thing is, or, in other words, the more energetic are its vibrations, the more complete, stable, and strong are the vibrations of the parts of which that thing is composed. The modern physicist tells us that the stones of which St. Paul's Cathedral is built, consist of millions upon millions of small particles called molecules ; and that although St. Paul's Cathedral seems to be absolutely at rest, as if it would last for ever, and although each particular stone seems equally so, yet when you get down into the intimate structure of each stone, and of every part of the fabric, you get nothing but a multitudinous ocean of motion. What appears to us solid and at rest, is absolutely in a perpetual state of unrest ; in fact, its stability consists in its state of unrest.

The difference between a source of light, such as a glowing solid or liquid, which, when analysed, gives us a continuous spectrum, and a gas or a vapour which does not give a continuous spectrum and which does not therefore give us white light, is simply this, that in the case of gases and vapours which are produced by the atom-dissociating power of electricity and of heat, those molecules which give us those coloured phenomena differ only from the larger ones, which give us a continuous spectrum, in that, owing to the action, upon the one hand of electricity and upon the other hand of heat, they are much simpler than the others.

As we melt a metal such as sodium, or even other metals of a very much more refractory nature, all of those metals which give us the beautiful rainbow band called the continuous spectrum, to start with, come at last to a stage at which the spectra consist of one, two, four, eight, or hundreds of lines, as the case may be. But there are between those stages other intermediate spectra, which seem to show us that as the action of electricity or of heat is allowed to go on, the particles, whatever they may be, of which matter is built up, and which give us white light when we get solids or liquids to radiate, really become more and more simple, until at last we get that line spectrum to which reference has been made.

Now let us apply this to those elementary substances which we can get at and experiment with, which are within our reach, and those also about which we know absolutely nothing, except what the spectroscope alone can tell us, I mean the materials of which every celestial body is made up.

In regard to elementary matter, we have first of all this fact, that if the particles under examination send us white light, we get a continuous spectrum from it ; therefore when we have to deal with white light, we know that we are dealing with matter in a solid, or liquid, or

densely gaseous state, but we do not know what matter
it is ; it may be any of the metals ; it may be any of
the compounds which will stand a high temperature, but
whether it is bismuth, or oxygen, or nitrogen, or lime we
do not know. But when we have got the matter simplified, so that its particles, instead of being complex enough
and self-contained enough to give us this white light,
are broken up, and give us coloured light, then we find
that no two substances with which we are acquainted
give us the same sets of lines.   Hence the origin of the
term Spectrum Analysis, as the study of the spectrum
thus enables us to tell one substance from another.

These coloured senders—these particles of matter
otherwise called molecules—that send out coloured light,
which being analyzed gives us these lines, that are really
and truly things infinitely small beyond our conception,
but yet absolutely and truly vibrating bodies, and the
spectrum is the result of the vibrations.

That idea leads us further, and it enables us to say
not only that such and such a spectrum is given by
such and such a substance, but also that such and such
a spectrum is given by that substance within a certain
range of temperature, while other conditions which will
be fully referred to in the sequel are not without their
influence.

Hence, with vapour as a sender out of light, we learn
from the spectrum its chemical composition, its density,
roughly its temperature.   The same vapour when, instead
of being used as a sender, it is used as a medium, gives
us exactly the same spectrum reversed, so that we can

detect the presence of sodium vapour, to take an example, when it is sending out light by means of its vibrations set in motion by heat, and when it is between us and any sender whatever which can feed it with those same vibrations ; and we have in both cases the same means of determining roughly that it is of a certain temperature, and that its density is within certain limits.

It is by following out considerations of this kind that all the stars in heaven have revealed to us their constitution—that is to say, the elements of which they are built up, at what temperature they exist, and a great deal of their meteorology : by which I mean the nature and extent of their atmospheres, and the way in which their atmospheres vary from cycle to cycle.

The accompanying coloured plate (Pl. II.) brings together representations of the various classes of spectra to which reference has been made in this chapter.

The first group of spectra, 1 to 8 inclusive, deals with the giving out of light—with what, in scientific language, is called radiation.   In radiation we are studying the vibrations of molecules conveyed to us by the ether, wave length for wave length.

The second group (9 and 10) deals with absorption. In absorption we are studying the molecular structure of bodies which are absorbing, and the etherial vibrations impinging on that structure are stopped or weakened, wave length for wave length.

The first group subdivides itself into two. In one we have the continuous spectrum—the complete series of light waves always seen when we are dealing with

Plate II

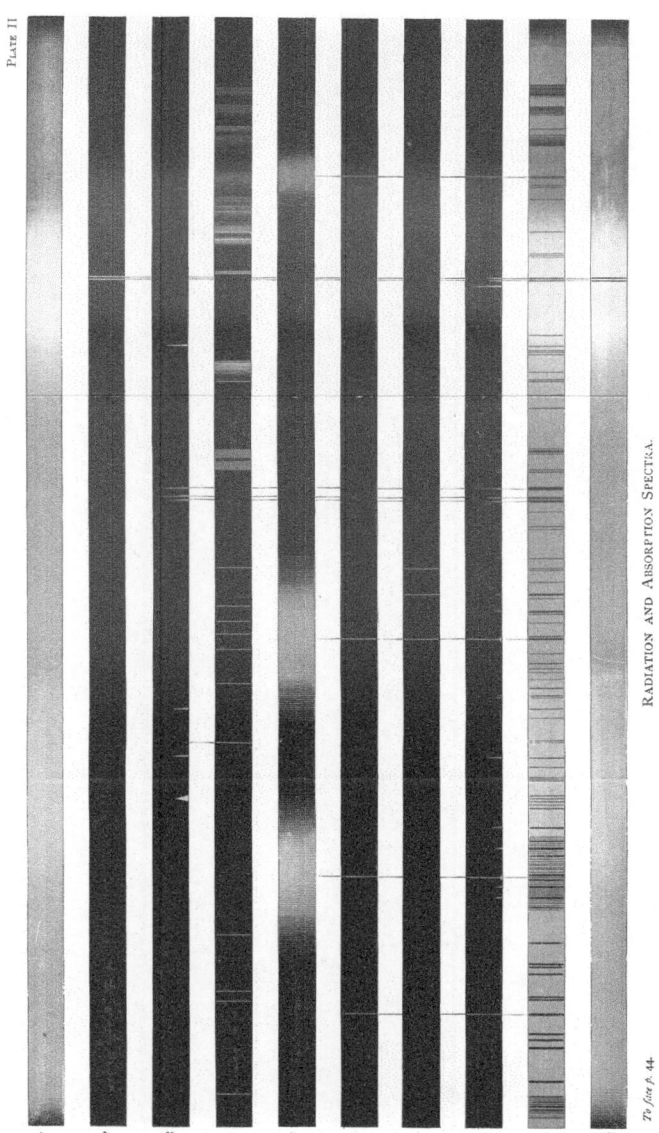

RADIATION AND ABSORPTION SPECTRA.

To face p. 44.

the giving out of light by a body, whether simple or compound, which retains the solid or liquid form while in a state of incandescence. A white-hot poker and glowing fluid iron are familiar instances of such a light source.

In 2–8 we have examples of so-called line or band spectra produced by vapours or gases giving out light. These require detailed examination. In 2, 3, we have spectra belonging to two elementary metallic bodies— sodium and magnesium. It will be seen how different they are, and how easy it is to bear in mind or to map such an allocation of lines, so that when produced from an unknown body the existence of either can be detected by such spectral examination. This part of the plate might have contained the spectra of all the metallic bodies in a state of vapour, and it would still have been abundantly apparent that each was so distinct from all the rest, that there would be little difficulty in recognizing the dissimilarity. In 4 we have a specimen of the spectrum of a compound body, in this case a salt of strontium—other salts have equally characteristic spectra. Here again is a complete separation from 2 and 3, and an equally complete dissimilarity is observed even in the spectra of compounds which ordinarily occur in the state of gas. In 5 and 6 we have bands of light occupying the same position in the spectrum, but they are wide in one case and narrow in the other. We are dealing with the same substance—hydrogen—at different pressures.

In 7 and 8 we have left the earth and have sought for

examples from the skies ; the spectra given are those of the nebulæ and of the Sun's chromosphere. The evidence on which the assertion is made that hydrogen exists in some of the nebulæ, and sodium, magnesium and hydrogen in the Sun's chromosphere, will be at once appreciated.

In 9 we have the absorption in the solar spectrum. In 10 the absorption of sodium vapour in the laboratory. The coincidence of the double lines in the yellow supplies the basis of solar and stellar chemistry.

In this plate I have limited myself for the sake of clearness to those spectra which contain lines. It will be seen in the sequel, however, that all bodies have another kind of spectrum which is dissimilar for each, called a *fluted* or *channelled space* spectrum. In the sequel attention will be fully called to this as well as to other spectra again, which, like the continuous one, are the same or nearly the same for many bodies and most probably for all. Here, however, we are entering somewhat into detail, and it will be as well to conclude the introductory chapter at this point, and proceed to discuss some of the experimental methods by which the phenomena are produced.

CHAPTER II.

METHODS OF DEMONSTRATION AND LABORATORY
WORK.

I HAVE had elsewhere to deal with the construction of the spectroscope,* I propose therefore chiefly in this chapter to deal with its use in the laboratory and the various methods of demonstrating spectrum phenomena.

It is as well, however, to state by way of reminder, that the spectroscope, however simple or complex it may be, is an instrument which allows us to observe the image of the slit through which the light enters it, in the most perfect manner, and this for every wave length of light. If the entering light contains rays of every wave length, then the images formed by each will be so close together that the spectrum will be continuous, that is without break. If the light contains only certain but not all wave lengths, either from defective radiation or owing to subsequent absorption, then we shall get certain, and not all, of the possible images of the slit, and the spectrum will be discontinuous.

* "The Spectroscope and its applications," by J. Norman Lockyer. Macmillans.

§ 1. *Use of the Lens.*

I will first call attention to the difference between the ordinary spectroscope and one that has a lens in front of the slit. The use of this is as follows : If we have an extremely complex light source, let us say a solid and a mixture of gases giving us light, and we allow the light to enter, so to speak, indiscriminately into the spectroscope, then in each part of the spectrum we shall get an integration of the light of the same wave length proceeding from all the different light waves. But if by means of a lens we form an image of the light source so that each particular part shall be impressed in its proper place on the slit plate, then in the spectrum the different kinds of light will be sorted out, and if there be any variety in the phenomena which these different sources of light present to us, they will be all clearly shown in the spectrum visible in the observing telescope, or in the photograph if we photograph the result.

There is a simple experiment which all can try for themselves, which shows clearly the different results obtained. If, for instance, we observe the light of a candle with the spectroscope in the ordinary manner, that is by placing the candle in front of the slit at some little distance from it, we see first of all, a band of colour —a continuous spectrum, and in one particular part of the band we see a yellow line, and occasionally in the green and in the blue parts of the band other lines are observable. Now, if we turn the spectroscope, which without a lens is an 'integrating' one, to adopt Professor Young's very convenient phraseology, into an 'analyzing' one, by *throwing an image of the candle on to the slit* —the slit being horizontal and the image of the candle vertical—we then get three perfectly distinct spectra. We find that the interior of the candle, that is the blue part, best observed at the bottom of the candle, gives us one spectrum, the white part gives us another, while on the outside of the candle, so faint as to be almost invisible to the eye, there is a region which gives us a perfectly distinct spectrum with a line in the yellow. In this way there is no difficulty whatever in determining the co-existence of three light-sources, each with its proper spectrum, in the light of a common candle.

The method of observing spectra then, to which I have referred, and which has been adopted in most of the work of which I shall give an account in the sequel, consists in throwing an image of the light-source on the slit of a spectroscope in laboratory experiments in exactly the same manner in which I proposed, in 1866, that an image of the sun should be thrown on the slit in order to spectroscopically examine minute portions of the sun and his surrounding atmosphere.

It is obvious that in this method the image of the slit will be associated in the spectroscope with an image of a section of the light source, and if this be a spark, and if from any cause there be various shells of vapour surrounding each pole, which shells give different spectra, then these spectra will be sorted out so that their variations may be traced from pole to pole.

E

This method was employed by Dr. Frankland and myself in our experiments in 1869, and was first ex-

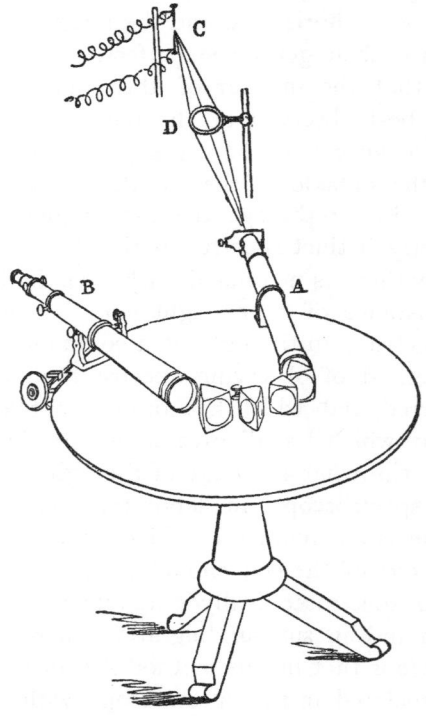

FIG. 14.—The analysing spectroscope
A. Collimator.   C. Spark.
B. Observing Telescope.   D. Lens.

hibited at a lecture at the Royal Institution, in 1870. The same method has more recently been employed with great success by M. Salet in a research on the spectra of the metalloids.

The arrangements adopted will be easily gathered from the annexed woodcut (Fig. 14) which represents a spectroscope with a lens in front of the slit. It is scarcely necessary to add that an important condition of this method is that the object-glass of the collimator should be filled with light, and also that no light should be wasted. So long as these conditions obtain, conjugate foci and different lenses may be employed and the size of the image varied at pleasure, and still the brightness of the spectrum will be sufficient in all ordinary cases.

We drive the metal of which the poles at C (Fig. 14) are composed into vapour, the vapour is rendered incandescent; the spectrum we get will therefore be one of bright lines. Now, when, instead of merely bringing the spectroscope opposite to the poles, in which case, in every part of the spectrum we should get light from every part of the spark, we use a lens, by means of which an image of the spark is thrown on the slit, then in each strip of the total visible spectrum is the spectrum of some particular part of the vapour. The poles are perpetually giving off vapour, which is cooling and going away; some of it is being oxidised, some of it is travelling away along the air-currents set up. What follows? There must be more vapour close to each pole than in the interval between the poles: that will be still more true if we make the interval between the two poles longer. In the region between the two poles,

E 2

if they consist of two different elements, we have three distinct spectra produced (Fig. 15).   In the upper part, a region rich in the lower vapour ; in the lower, one rich in the upper vapour ; between them one which is rich in neither.   We have then at least three distinct layers, so to speak, in the spectrum : the spectrum of the vapour of the upper pole, the spectrum of the vapour

Violet   Red

FIG. 44.—Copy of a photograph of the spectrum of the spark between poles of zinc and cadmium, showing the separation of the three spectra.

of the lower one, and also of the central region.   The number of particles of each vapour will decrease from each pole.

We see in a moment that much the same condition of affairs will be brought about, if, instead of using a spark, we use an electric arc, in which the pure vapour of the substance which is being rendered incandescent fills the whole interval between the poles, the number of particles being smaller at the *sides* of the arc.   Now we can throw an image of such a *horizontal* arc on a vertical slit ; the slit will give then the spectrum of a section of the arc at right angles to its

length. In Fig. 17 we have a reproduction of a photograph of such a spectrum. I wish to draw attention
to the long and the short lines. The vapour which
exists furthest from the core of the arc has a much more
simple spectrum than that of the core of the arc itself.

Fig. 16.—Image of horizontal arc of sodium vapour thrown on a vertical slit
(from a photograph).

The spectrum of the core consists of a large number
of lines, all of which die out until that of the part of
it furthest from the centre consists of one line.

In order to demonstrate this difference in the phenomena observed, let us say, in the candle experiment,

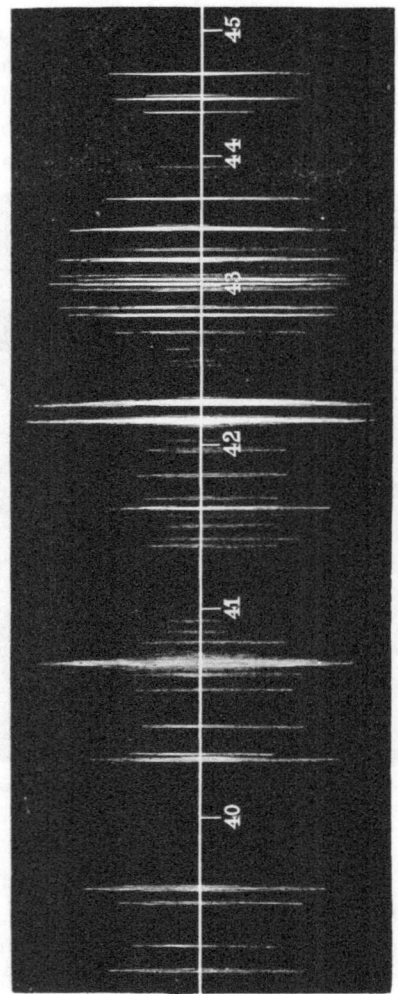

FIG. 17.—Long and short lines.

The illustration given is a copy of a photograph taken when the salt of strontium, strongly impregnated with calcium, is made to vaporize between the carbon poles in an electric lamp fixed in a horizontal position. The two longest lines, between W. L. 42 and 43, are due to strontium and calcium; the longest line, between 40 and 41, is due to strontium; the four short lines to the left are due to impurities of calcium and aluminium in the carbon poles.

by means of an electric lamp ; we first use the poles as
a source of white light, in the ordinary way, and insert

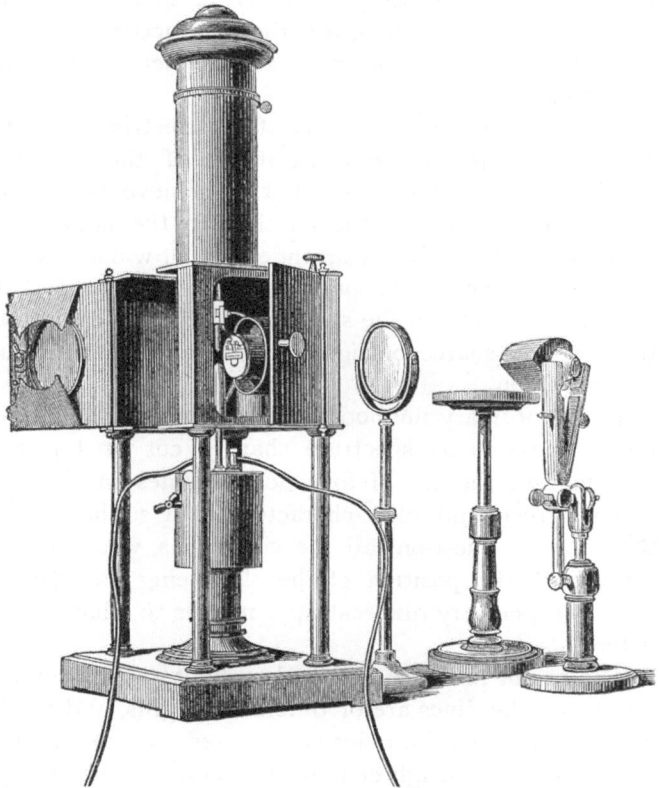

FIG. 18.—Arrangement of electric lamp for demonstrating the existence of long and
short lines.

in the lower one a particular substance which will give

us the bright yellow line already referred to, the result will be that we shall get a mixed light-source, and therefore a mixed spectrum, in which these two lights will be mixed if we integrate the light according to the ordinary method ; a different arrangement is however possible.

In demonstrating by means of the electric light it is not convenient to throw the image of the points of the electric lamp upon a slit, but we achieve nearly the same result if we place the slit close to the points. In Fig. 18 we have such an arrangement, in which the slit is almost touching the source of light, so that if there are any variations in the spectrum of the different portions of the source of light we get a different result on the screen.

By the ordinary method, when we employ sodium we have a continuous spectrum sharply cut at top and bottom and the set of four double lines in the red, yellow, green, and blue characteristic of sodium : with that now in question, all the conditions, with the exception of the position of the slit, being exactly the same, we get very different appearances to what we got before.

The bright lines are obtained without any continuous spectrum, the lines are of different lengths, and as the distance between the poles is increased, one line is seen alone and much brighter than all the rest. The yellow line in the case of sodium will remain after the other lines have considerably dimmed or even disappeared.

This is one example of the kind of difference in the

phenomena observed when the new method of demonstration is adopted.

The spectroscope thus armed is in fact a new instrument.

§ 2. *Radiation Phenomena.*

In discussing the various methods of using the spectroscope the subject may be conveniently divided into two perfectly distinct parts, the first dealing with the radiation of light, and the second with absorption. Now in order to get radiation from any substance whatever, we have to set that substance in vibration in some way or other. If we are dealing with a certain class of substances—I shall state what class by and by —we find that the vibration which is necessary for radiation is conveniently set up by the employment of heat. In other cases, however, heat utterly fails us, and we have then, as stated elsewhere, to resort to electricity ; and what heat does for us at one end of the scale, electricity does with equal convenience at the other.

The instrument generally adopted for showing the action of heat upon the salts of most of the metals is the Bunsen burner, placed, with or without a lens, in front of the slit of the spectroscope ; the spectral phenomena thus observed unfortunately cannot be thrown on the screen by means of the electric lamp, because electricity is so different in its action to heat, that if we attempt to employ the electric lamp with these salts we at once get totally different phenomena.

Another class of apparatus is necessary when we

wish to employ no longer heat but electricity as the source of the vibrations, in one case of solid substances, in the other of gaseous bodies. In the case of the former by means of an induction coil a spark is made to pass between various metals, conveniently arranged on a holder of a circular form, round the rim of which is placed a series of clips, which by rotation can be successively brought under a fixed upper one. This can be managed in such a way that the current is allowed to pass when the fixed and movable electrode are in the same vertical line in front of the slit of the spectroscope. The very great convenience of this arrangement is, that before a series of experiments is made, each of these particular holders may be charged with the particular metals it is wished to observe ; and then they may be rotated in succession in front of the slit of the spectroscope. In order that the spectra may be best seen it will be necessary to regulate the distance of the electrodes apart, and if a jar be used, it should be of such a size as to give the greatest constancy to the spark.

The following method*is a convenient one for observing the spark spectra of salts. Some pieces of stout aluminium wire ten millims. long and three millims. in diameter are taken ; one end is flattened for about one third of the length for the purpose of inserting it in the spark-holder, and the other drilled down in the direction of the axis for from two to three millims, thus forming a small conical cup; a very fine hole is drilled

* This method was devised by Mr. Friswell.

through the side of this cup at the bottom and the flattened end carefully split. Through the lateral hole a piece of platinum wire 0·5 millim. in diameter is passed,

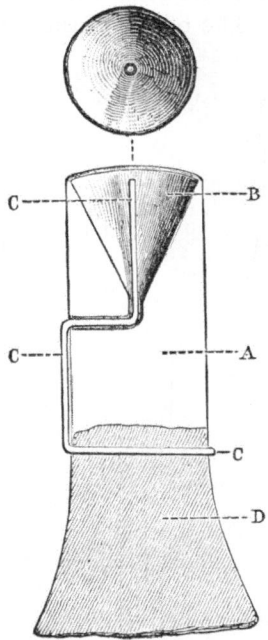

FIG. 19.—Plan and section of cup used with salts. A, aluminum wire; B, cup-shaped cavity drilled in it; C, platinum wire; D, flattened and split portion of the aluminium wire.

and one end brought round through the split end of the aluminium, the other being brought up the centre of the cup. The split is now closed by strong pressure in

a vice, and the ends of the platinum wire cut off.   The
whole now presents the appearance of a small candle,
the platinum wire representing the wick : the accompanying figures (Figs. 19 and 20) will render the preceding
statement clear.   The object of the "wick" is to confine

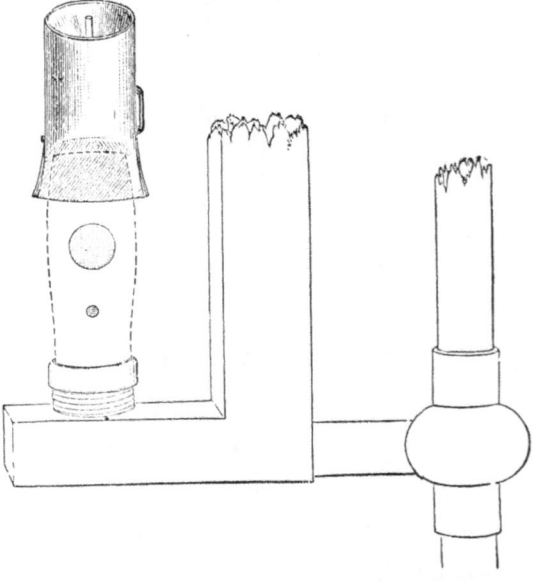

Fig. 20.—Aluminium cup placed in the spark-stand as in use.

the spark to the centre of the dry salt.   Without it the
spark is very unsteady, leaping about from side to side
of the cup.   Round this wick the salt in fine powder is
tightly rammed down.

A similar cup, without the wick, may be used for the examination of the spectra of metallic barium, strontium, and lithium, &c., the metal being hammered into it.

One of these cups with the salt replaces the lower pole in the spark-holder, the upper-one may be composed of copper, that metal being a good conductor and giving a very simple and easily recognized spectrum.

When it is desired to observe the spectrum of salts in gases other than air, the salts are rammed into the small aluminium cups, and the cups are fastened to copper rods passed through a cork into the interior of a wide glass tube. An aluminium point is used for the opposite pole, which is also fastened to a copper rod passing through a cork fitted into the other end of the tube. Both corks are pierced and furnished with narrow glass tubes ; one of these serves to admit the gas, while the opposite one acts as an exit-tube.

The gas properly purified and dried is admitted in a gentle stream into the tube containing the poles, and the spark is then passed.

In the examination of the spectra of gases a revolving holder may also be introduced. A series of Geissler's tubes, containing different gases at very low pressure is so mounted that as each one, when the stand is rotated, comes before the slit, it is in contact with a metallic point, and the gas is rendered luminous, so that it is possible in a very short time to make a considerable number of observations or comparisons. For observations of gases at ordinary pressures the electrodes, with the power of varying the distance between them, are arranged in a tube, through which we can determine a flow of any particular gas we wish to examine.

In addition to the observation of vapours and gases, it is necessary oftentimes to make observations of the spectra of solutions.

There are many ways of accomplishing this; the solution to be examined is put into a small cup, so that the solution is just flush with the top of a platinum point fixed into the bottom of the cup. There is another platinum point connected with the other terminal above, and in this way the spectrum of the solution may be conveniently observed. There is, however, in this method of working the great objection that the quantity of solution between the poles is perpetually varying, and two of M. Dumas's students, MM. Delachanal and Mermet, have quite recently suggested an extremely interesting modification. They use an ordinary test-tube with a platinum terminal soldered into the bottom, and a cork at the top through which a glass tube pierced by a platinum wire is allowed to fall. The special arrangement introduced is this: there is a fine bit of glass tubing dropped on the lower terminal, and the height of the solution is so arranged that by capillarity a certain portion of it rises between the lower terminal and the glass into the central cavity and fills the little cup. The moment the cup is overfull the solution flows down the side, and in that way a most perfect standard is obtained.

The following advantages are claimed for this method—

1. Constancy of spark permitting prolonged observa-

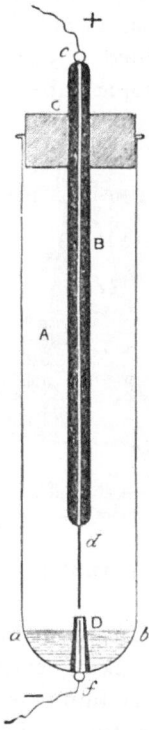

FIG. 21.—A, tube, 11 mm. high, into which the liquid to be analysed is poured ; B, capillary tube in which is fixed the platinum wire *c d*, which forms the upper electrode ; C, cork stopper closing the tube A—it supports B, and permits its being moved with little friction ; D, small capillary tube, slightly conical, covering the lower electrode *f*; *d*, upper electrode ; *f*, lower electrode ; *a b*, level of liquid.

tion of spectra.   2. Suppression of the meniscus, and
consequently of the absorption which it produces by
partly concealing the spark.   3. Electrodes enclosed in
a special tube, which preserves the solution from contact with impurities.   4. Possibility of collecting entirely
the substance examined.   5. Possibility of arranging
a series of spectroscopic tubes, enclosing solutions of

Fig. 22.—Examination of the spectra of solution by blowing the solute into the Bunsen
flame by means of a "spray inhaler."

various bodies, thus permitting rapid demonstrations
and comparisons.

   To work the apparatus, pour into the tube A the solution to be examined, taking care that the electrode f
and the tube D are only immersed to half their height.
Let $a\ b$ be the level of the liquid; capillary force determines the ascent as far as the point D, on which is
formed an immovable drop which is vaporised when an
induction current is put on by c and f.   The observa-

tions may then continue a very long time without intermission, allowing the spectra to be observed and drawn with the greatest ease.

There is yet another very convenient method of observing the spectra of solutions which I have recently adopted. Take one of those very convenient little steam jets much used by physicians. (Fig. 22.) We have water in the upper part which can only escape by a small aperture and a little burner below which makes the water boil, and we get a steam jet playing over a vertical shaft of glass, the lower end of which is immersed in a bottle containing the solution one wishes to observe. When the water boils, we have first of all the jet of steam by itself, and the effect of that will be to exhaust the tube ; the effect of the exhaustion will be to suck the solution up the tube to the very narrow aperture at the top, and when that is reached we get the solution in the finest possible spray, which is always the most favourable condition for observing phenomena of this kind. When the steam jet is at work and allowed to play on a Bunsen flame, the flame is throughout its whole length impregnated with the salt in solution.

It is unfortunate that it is impossible to demonstrate these phenomena to a large class.

If we take the salts and expose them to the action of electricity in the lamp, we obtain the typical spectra of the *metals* themselves, and not of the *salts*, for a reason which will be stated further on.

F

§ 3. *Absorption Phenomena.*

When we pass from the general phenomena of radiation, the different methods of obtaining light sources, and of observing their spectra, to the other large branch of work, the phenomena of absorption, we come upon a ground in which the experimental methods are entirely changed. In studying the phenomena of absorption what we have to do is no longer to observe the different kinds of light given out by bodies, but to observe the action of different bodies on light containing rays of all visible refrangibilities.

Now, the light which one can use in small laboratory experiments for these extremely interesting researches may be the light of the common candle, or of gas, or of anything which gives us white light, but teachers, for the purposes of demonstration, prefer to use the brightest white light, namely, that of the electric lamp.

What one has to do in studying the absorption of different bodies is to immerse them in the light proceeding from the particular light source we choose to employ, whether it be a candle, gas, or the electric lamp.

As long as a body is solid, one simply has, as in the case of differently coloured glasses, for example, to put it in front of the slit of the spectroscope, and therefore for demonstration in front of the slit of the lamp, and observe what particular light is stopped by the material. But, as in the case of radiation, we are not limited in our inquiries on absorption to solid bodies,

we wish to observe what happens when we allow light
to pass through vapours, or gases, or liquids. Different

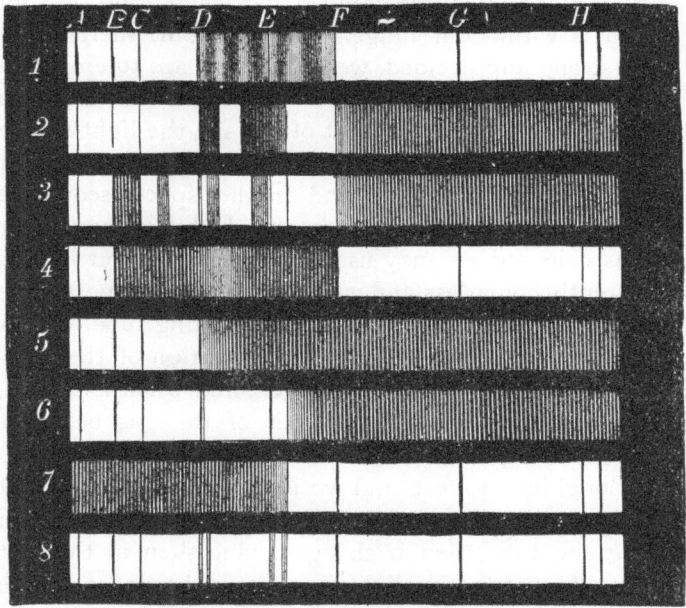

FIG. 23.—Diagram illustrating the various kinds of absorption, and the characteristic
absorption of some bodies. 7, Absorption of the red end of the spectrum; 5 & 6,
absorption of the blue end; 1, special absorption of permanganate of potash; 2, ditto
of blood; 3, ditto of chlorophyll; 4, ditto of cobalt glass (the characteristic structure is
not given); 8, ditto of didymium glass.

arrangements may be employed. Let us take, for
instance, the case of a vapour. We need not discuss the
question of the permanent gases because they are colourless; that is, their absorption is out of our range of

F 2

vision, but in the case of vapours some of them have a very definite colour indeed, the coloration of the vapour meaning that it is absorbing a particular kind of light which is within our range. If the colour, therefore, is very strong and decided, we must take care to employ a small thickness only, and it is convenient to employ a tube. This is placed in front of the slit, the light source being behind it, so that a constant beam of white light passes fairly through it ; and in the spectroscope we observe what kind of light is absorbed. If the vapour is less coloured we may use a longer tube, and if it is apparently colourless and we have reason to expect some absorption somewhere, we use a very long tube. Thus I have prepared a tube for the observation of the spectrum of ozone mixed with oxygen, and when it is used, light will pass through 120 feet of the gas before it reaches the spectroscope. The tube itself is not 120 feet long, but 40 feet, and we use two reflectors, one at each end, to make the light go first in one direction along the tube, then back again, and then in the first one again, until it is allowed to enter the spectroscope after three transmissions along the tube.

Such tubes are very convenient for the observation of vapours at the ordinary temperature, but it would not do to limit our observations to this temperature, because a great many substances give off no vapour at anything like ordinary temperatures. When one wishes to observe the vapours, say of sodium, platinum, iron or any metal in fact, it is necessary to employ high temperatures—in some cases very high temperatures.

For that purpose a furnace is employed into which is inserted an iron tube ; the tube is kept filled with hydrogen, and the metal to be examined is inserted by means of a side tube when the tube is red hot or white

FIG. 24.—Absorption of different thicknesses of the colouring matter of litmus.

hot. The metal slides down this tube and falls into the white-hot portion ; an electric lamp is placed at one end, and a spectroscope at the other, and in that way the absorption of vapours at moderately high temperatures can be conveniently observed.

The iron tube is four feet in length, and is provided with a central enlargement, suggested to me by Professor Dewar, forming a T-piece by the screwing in of a side tube, the end of which is left projecting from the door in the roof of the furnace. Caps are screwed on at each end of the main tube ; these caps are closed by a glass plate at one end, and have each a small side tube for the purpose of passing hydrogen or other gases through the hot tube. The furnace is supplied with coke or charcoal. The temperatures reached by this furnace may be conveniently divided into four stages :—

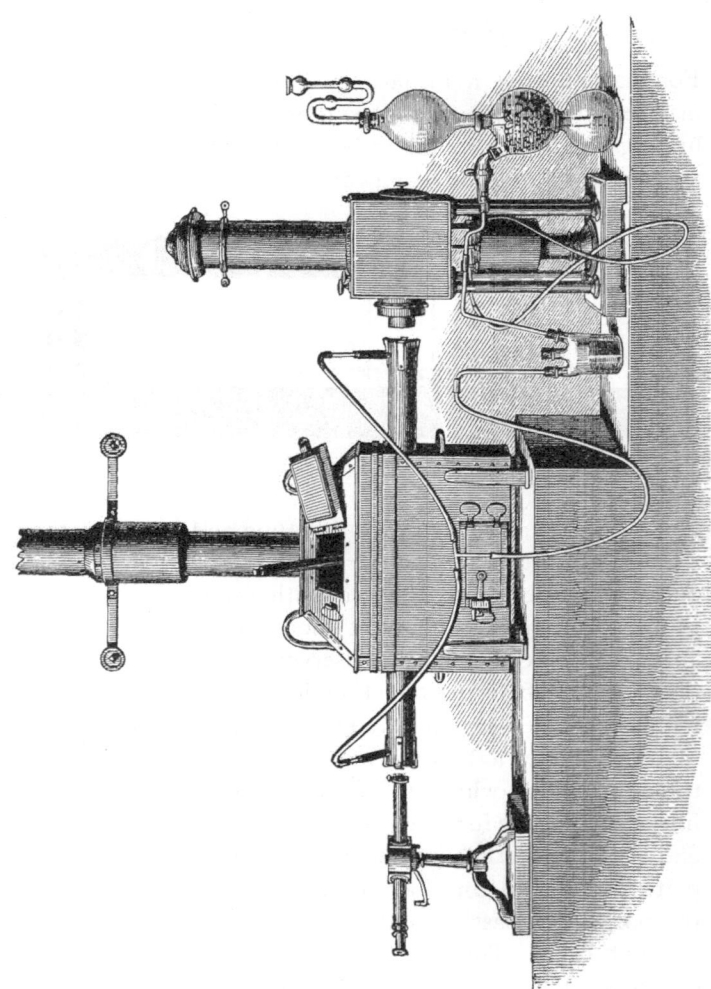

FIG. 25.—Arrangement for observing the absorption spectra of metallic vapours.

I. When the continuous spectrum of the tube extends to the sodium-line D, this line not being visible.

II. When the continuous spectrum extends a little beyond D, this line being visible as a bright line.

III. When the spectrum extends into the green, D being very bright.

IV. When the spectrum extends beyond the green and D becomes invisible as a line, and the sides of the furnace are at a red heat.

There is one matter to be attended to in this branch of work when we have to deal with such metals as sodium and potassium. These metals are melted gently under paraffin, and when in a liquid state are drawn up into fine glass tubes—a method proposed by Professor Dewar. When we wish to observe the spectrum, a portion of the tube with the included metal is broken off. The tube not only protects the metal from oxydation but forms a sort of carriage on which the metal slides down the inclined adit into the central cavity of the tube without any difficulty. In the absence of such an arrangement the metal would stick in the side tube, melt and volatilize, instead of getting down to the bottom and giving off its vapours in the cavity.

If a still higher temperature is required, then the ordinary furnace is replaced by one competent to give a higher temperature still; the iron tube is replaced by one of chalk, and the metals are driven into vapour by means of the oxy-hydrogen blow-pipe, taking care to have an excess of hydrogen.

This instrument, devised by Sainte-Claire Deville and Debray,* renders it possible to attain high temperatures with great facility, and Stas has employed their method in the distillation of silver.† The lime still arranged by him was modified in that about to be described, employed by Mr. Roberts and myself, in order that the metallic vapour might be conducted into a lime tube or tunnel heated to whiteness, so placed that a beam from a lamp could traverse it.

The apparatus employed in such researches as these is shown in figure 26, in which A is the block of lime‡ divided horizontally by a plane through the axis of the tube (B B'), this tube being 16 centims. long and 30 millims. diameter. The receptacle (C) communicates with the centre of B B', and is open at the upper surface of the lime block, in order to admit of the introduction of the oxyhydrogen blowpipe (D), which is provided with a thick nozzle of platinum 20 millims. in diameter. The ends of the tunnel in the lime were closed by glass plates held on by a suitable clip. Small lateral orifices were cut in the lime for the insertion of tobacco-pipe stems, through which a stream of hydrogen could be passed into the tube and receptacle.

An electric lamp (F), in connection with a 30-cell

---

* Ann. de Chimie et de Physique, tom. lvi. p. 413.

† Stas, 'Sur les lois des proportions chimiques, p. 56.

‡ We were indebted to the well-known metallurgist, Mr. John S. Sellon, of the firm of Johnson and Matthey, for a pure variety of limestone from which the blocks were prepared, and it answered its purpose admirably.

Bunsen's battery, was placed opposite one end of the tube, and a spectroscope (G) opposite the other. (Fig. 26.)

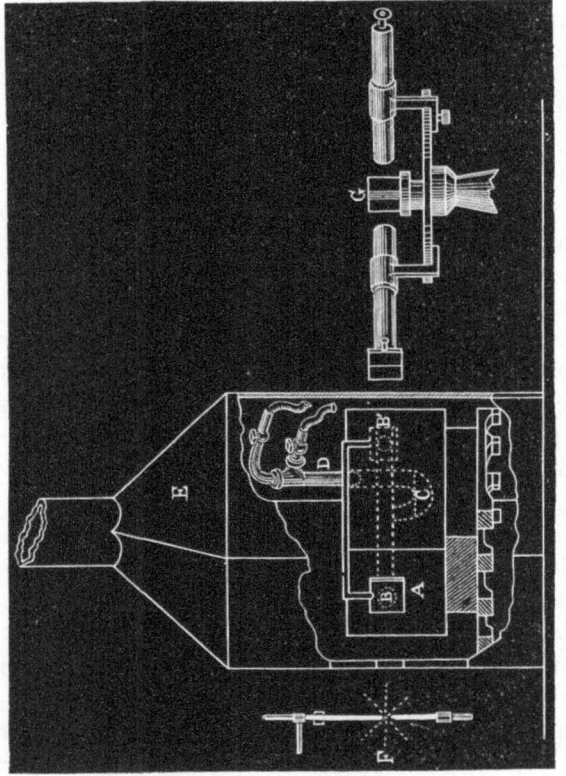

Fig. 26.—Arrangement for observing the absorption spectra of the metals driven into vapour by the oxyhydrogen flame.

When it is desirable to increase the length of the column of vapour, a tube some centimetres long is made in a fresh block of lime, the cavity being arranged as before ; in each end a short accurately fitting iron tube, luted with a mixture of graphite and fireclay, is inserted ; the total length of the column may thus conveniently reach 60 centims.

The lime block (C) with its fittings is placed in the charcoal-furnace (E), by means of which the whole is raised to a high temperature. As soon as the block is heated to bright redness, the metal, the vapour of which is to be examined, is introduced into the cavity (C), and the flame of the oxyhydrogen blowpipe (D) is allowed to play on its upper surface, care being taken to employ an excess of hydrogen. In almost every case the metal experimented on has thus been rapidly volatilized (the exceptions being gold and palladium). The central portion of the lime block was raised to a white heat by the action of the blowpipe in our experiments. As the glass plates rapidly become clouded by the condensation of the metallic vapours, it is necessary to adopt an arrangement by which they can easily replaced.

Among the precautions adopted in order to assure ourselves that oxides were not present to disturb the accuracy of the results, one of the glass plates was removed at the conclusion of each experiment, and the presence of an excess of hydrogen in the tube, conclusively proved by igniting it at the open end.

We were enabled at any time, by modifying the conditions of the gas-supply, to introduce the spectrum of the oxyhydrogen flame. It may further be stated that, with few exceptions, the metals were previously melted in a stream of hydrogen and enclosed, until experimented on, in sealed glass tubes.

There is still another case, that of liquids, to consider. Many different forms and constructions of what are called *cells*, to hold liquids, have been devised. There is only this to be said about them in this place, that it is very important to know exactly what composition has been employed in cementing the glass together, and, as a matter of fact, for certain solutions one must employ certain cements.

Many of the phenomena of absorption may be demonstrated to a class by means of the electric lamp. We require first a pure continuous spectrum, and having it on the screen the different substances into the nature of which we wish to examine are placed between the slit and the lens.

It is important to be able to observe the varying effects of pressure and density upon spectral phenomena. The following method of observing this in the case of sodium vapour was devised by Dr. Frankland and myself in 1869 :—

(1) Into a piece of hard glass combustion-tube, thoroughly cleaned and closed at one end, a few pieces of metallic sodium, clean and as free as possible from naphtha, were introduced. The end of the tube was then drawn out and connected with a Sprengel pump and exhausted as rapidly as possible. Hydrogen was then admitted, and the tube re-exhausted and, when

the pressure was again reduced to a few millimetres, carefully sealed up. The tube thus prepared was placed between the slit plate of a spectroscope and a source of light giving a continuous spectrum.

Generally, unless the atmosphere of the laboratory was very still and free from dust, the two bright D lines could be distinctly seen on the background of the bright continuous spectrum.

The tube containing the sodium was then heated with a Bunsen flame and the spectrum carefully watched. Soon after the application of the heat, a dark line, thin and delicate as a spider's thread, was observed to be slowly creeping down each of the bright sodium lines and exactly occupying the centre of each. Next, this thin black line was observed to thicken at the *top*, where the spectrum of the *lower* denser vapours was observed, and to advance downwards along the D line, until arriving at the bottom they both became black throughout ; and if now the heat was still applied, thus increasing the density of the various layers of the sodium vapour, the lines began to broaden until, in spite of considerable dispersion, the two lines blended into one. The source of heat being now removed, the same changes occurred in inverse order ; the broad band split into two lines, gradually the black thread alone was left, and finally that vanished, and the two bright lines were restored.

(2) This experiment was then varied in the following way. Some pieces of metallic sodium were introduced into a test-tube, and a long glass tube conveying coal-

gas passed to the bottom, an exit for the gas being also provided at the top. The sodium was now heated and the flow of coal-gas stopped. In a short time the reversal of the D lines was complete. The gas was now admitted, and a small quantity only had passed when the black lines were reduced to threads.

In studying the same phenomena in the case of the permanent gases, the aid of the Sprengel pump is indispensable. In this pump the pressure in a system of glass tubes is reduced by the fall of a broken stream of mercury ; a portion of the gas gets entangled between each drop, and thus escapes.

The gas properly purified, and if necessary dried, is introduced into the apparatus to which is also attached a Geissler tube provided with electrodes, for it must be remembered that no source of what is ordinarily known as heat is competent to render any of the permanent gases, except hydrogen, incandescent. In the tubes there is ordinarily a capillary portion, and it is to this that the spectroscope is directed, as here the phenomena are more luminous than elsewhere. It is generally more convenient to commence operations at high pressures. An important adjunct to this pump is a barometer which enables the pressure at which any special phenomena are observed to be determined. For minute differences of pressure, especially near the vacuum point, Professor M'Leod has furnished us with a pressure-guage so delicate that pressure can be accurately estimated to the thousandth of a millimetre.

CHAPTER III.

ON SPECTRUM PHOTOGRAPHY.

§ 1. *Nièpce, Herschel, and Draper.*

IN the year 1839 Nièpce had brought photography to a more practical realization than it had been by any of his predecessors. He had then for some years allied himself with Daguerre, and the daguerreotype was already in existence. The action of iodine on silver, first discovered by Fox Talbot, had been fixed by the vapour of mercury.* Now, in the daguerreotype we had not the action of light in its ordinary sense; and men's minds were very much exercised as to what could be the real cause of the effects which were then being revealed. A beam of light fell on a plate. On this plate was a certain chemical compound. What part of the sunlight, or was it sunlight at all, which so acted upon this compound that an image more or less permanent was obtained?

What more natural than that this question should be investigated by means of various tinted glasses? The solar beam which the experimenters then used they made to pass through glass, now of one colour, and now of another. There was an immense deal of

---

* Fox Talbot, *Philosophical Magazine*, vol. xxii. p. 97.

difference of opinion concerning the action of light as investigated in this way. In fact, I shall have shortly to show that Mr. Claudet and a French physicist, M. Becquerel, were considerably at variance with regard to one particular point which came out from this kind of investigation.

But we had not long to wait. Sir J. Herschel, in 1839, pointed out that it was not a question of investigating these new qualities of light at all by means of coloured glasses ; they should be investigated by means of the spectrum. In three papers, communicated to the Royal Society in the years 1839, 1840, and 1842, he showed that the only really philosophic way of investigating this problem was by obtaining a pure spectrum, in which we have, at once, in different parts, something similar to what we get at different times when we deal with red glass, yellow glass, orange glass, green glass, blue glass, and so on. And having such a spectrum as this to deal with, and supposing such a spectrum thrown on to the photographic plate, it is quite clear that if there were something magical or unknown in the red rays which gave us this new action on the molecules of the particular chemical compound employed, or whether this magic really resided in the blue rays, that we should at once have this demonstrated in the most unmistakable manner, by action in the part of the plate on which the red rays fell, or in the part of the plate on which the blue rays fell.

Now, although Sir John Herschel was the first, in this country, to point out the extreme importance of this point of view, he was by no means the only one. Then, as now, there were distinguished Americans who were well to the front, and among them was Dr. Draper, the father of another Dr. Draper whom I shall have to speak of by and by. Those familiar with the enormous step in advance which was taken in spectroscopic investigations by Wollaston, who substituted a slit for a round hole, will be surprised to find that the first observations were conducted by throwing a converging beam of sunlight, giving an achromatic image of the sun, on the plate, through a prism. This method of procedure, of course, did not go so far as a better one might have gone, but it went a considerable way. Sir J. Herschel, from his observations made in this manner, stated that he had found a new kind of light—a new prismatic colour, "lavender grey," altogether beyond the blue end of the spectrum—altogether beyond the *blue* end of the spectrum, not the *red* end. Prof. Draper, on his part, also came in the main to the same conclusion, stating that he had discovered a "latent light."

§ 2. *The First True Spectra.*

When we come from the year 1839 to the years 1842 and 1843, we find a great advance—an advance as great, as far as photography goes, as Wollaston's advance on Newton was with regard to spectroscopic observation. Both Becquerel and Draper introduced, instead of this achromatic image of the sun, the simple arrangement of throwing sunlight through a slit and a proper combination of lenses and prisms on to a plate. The result was that on June 13, 1842, Becquerel did what I may venture to call a stupendous feat. He did what has never been done since, so far as I know.* He photographed the whole solar spectrum with nearly all the lines registered by the hand and eye of Fraunhofer. I do not mean merely the blue end of the spectrum, but the complete spectrum, from the " latent light "—the ultra violet rays of Draper—to the extreme red end. (Fig. 27.)

Draper also did something like the same thing, but not quite the same thing, in what he called a " tithono-graphic representation of the solar spectrum." He gives certain lines in the extreme visible blue part of the spectrum,† certain other lines, which none but Becquerel had ever seen before (Draper's work being nearly a year later), and in the extreme red—beyond the visible red of the spectrum—he gives other lines which even Becquerel had not photographed. This of course was a tremendous revelation.

A considerable discussion arose. Becquerel found, from an absolute comparison between the Fraunhofer lines which he had photographed and the Fraunhofer lines which Fraunhofer himself had registered, evidence in favour of the fact that this new chemical agent which was astonishing the world, whatever it was, was not something absolutely and completely independent of the visible rays. Draper, on the other hand, in

---

* " Bibliothèque universelle de Genève," t. xxxix.-xl., 1842, p. 341.
† *Philosophical Magazine*, vol. xxii. p. 360, 1843. For his earliest work see Journal of the Franklin Institute for the year 1837.

his "tithonographic representation," had, for some photographic reason or other, not succeeded in registering the
lines of the yellow, orange, and green parts of the spectrum, although he had fixed the lines in the blue, in the
extreme violet and in the extreme red ; and he considered himself justified by his experiments in coming
to exactly the opposite conclusion to that at which Becquerel had arrived, namely, that the light, whatever
kind of light it might be, which was at work in effecting
this chemical change which rendered photography possible, was something absolutely and completely independent of the ordinary light which the retina receives.

This was in the year 1843. By the year 1845, further
investigations by means of the spectrum had shown
that Dr. Draper's idea was heretical ; at the present day
it is the general opinion of physicists, that the radiations from any light source, from the extreme violet to
the extreme red, differ only in the rate and in the magnitude of the vibrations which are at work, so that I
claim for the application of photography to spectroscopy,
that it was not without influence in establishing the great
fact, that the visible, the chemical, and the heat rays
are really part and parcel of the same thing, that thing
being a system of undulations varying in rate and wave-
length from one end of the spectrum to the other,
whether we consider the visible portion or the invisible
rays—those outside the blue in one case, and outside the
red in the other. But this is not all : the application of
photography to spectroscopy has led to another important result. Sir J. Herschel, as soon as he applied the

prism, saw that it was no longer possible to proceed with that branch of research under the best possible conditions, unless opticians would construct lenses which would bring the visible and the chemical rays into absolute coincidence. This is now done by our Rosses and Dallmeyers in all camera-lenses.

§ 3. *The Ultra-Violet Spectrum.*

The next step brings us down to the year 1852. In this year a paper * was communicated to the Royal Society, by Prof. Stokes, who had already announced his discovery of what has since been called *fluorescence;* "on the long spectrum of the electric light." Prof. Stokes dealt in his first paper with the "change of refrangibility," or, as Sir William Thomson proposed to call it, "degradation of light," by virtue of which, light, which was generally invisible to us, could, under certain circumstances, be made visible. It is no part of my present purpose to go into this important paper at any great length ; but it may be pointed out that, as the *degradation* of light was in question, the invisible light to which Prof. Stokes referred as being capable of being rendered visible, must have been light outside the blue end of the spectrum, and not outside the red. Prof. Stokes, in his investigations, in order to get at this invisible light under better conditions if possible, than those with which he commenced operations, tested the transparency of the substances

* *Philosophical Transactions,* vol. clxii., 1852.

FIG. 27.—Reduced copy of Becquerel's photograph of the complete solar spectrum taken in 1842.

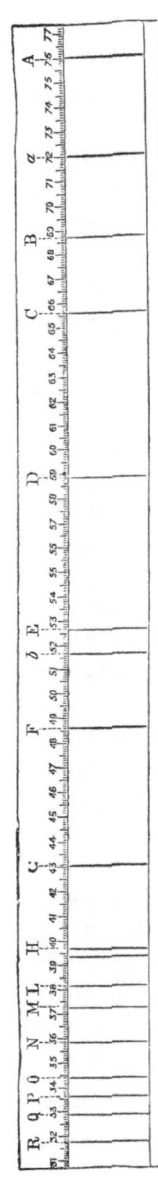

FIG. 28.—Wave-length solar spectrum showing the lines (from L to R) the positions of which have been determined by Mascart.

through which the light with which he experimented passed, and the transparency of glass was passed under review by him,* when he found that this invisible light, or whatever it was, could only get through glass with extreme difficulty. Continuing his investigations, he found that quartz on the other hand allowed this invisible light to pass. After referring to these experiments on glass and quartz, he proceeds to say : † —"I have little doubt that the solar spectrum" [which had already been photographed to a certain extent both by Becquerel and Draper beyond the visible blue end of the spectrum], "would be prolonged, though to what extent I am unable to say, by using a complete optical train, in every member of which glass was replaced by quartz." He then adds that other substances which suggested themselves to him were not equally good. Then further, that if this invisible light does get through quartz, and does become visible to the eye, it does not at all follow that it will be capable of being photographed. Because already Prof. Stokes, in order to continue his researches in fluorescence, had been, as it were, driven to photograph some of the results which he had thus obtained. I am sorry to say that, so far as I can find out, none of those photographs have ever been published.

In a note to his paper communicated to the Royal Society, he shows that his anticipations, so far as the eye was concerned, were perfectly justified by the facts.‡ He says :—"I have since ordered a complete train of quartz, of which a considerable portion, com-

* Op. cit. art. 202. † Art. 204. ‡ Page 559.

prising, among other things, two very fine prisms, has
been already executed for me by Mr. Darker ; with these
I have seen the lines of the solar spectrum to a distance
beyond H, more than double that of *p*. So that the
spectrum, reckoned from H (the outside line in the
portion originally visible), was more than double the
length of the part previously known from photographic
impressions.   The eye generally can see the two dark

H 2   H 1

FIG. 29.—The H-lines in the blue end of the solar spectrum, from a photograph by
the author.

bands represented in the middleof Fig. 29, and lettered
H 1 and H 2.   The least refrangible part of the spectrum lies to the right.   When Prof. Stokes therefore
stated that the solar spectrum was prolonged, he meant
that the part of the spectrum visible either to the
unassisted eye or on a photographic plate after impression, extends to a certain distance to the left of
these two dark lines.   The part which Prof. Stokes
rendered visible by means of his quartz train extended
a considerable distance to the left, beyond the part of
the spectrum represented in the figure.

So much for the solar spectrum. Prof. Stokes, in

a paper communicated to the Royal Society in 1862,* refers to his former paper, and to what he had been enabled to do by means of it. He states : " A map of the new lines " [the lines thus observed by him] " was exhibited in an evening lecture before the British Association, at their meeting in Belfast in the autumn of the same year, and I then stated that I conceived we had obtained evidence that the limit of the solar spectrum in the more refrangible direction had been reached. In fact, the very same arrangement which revealed, by means of fluorescence, the existence of what were evidently rays of higher refrangibility coming from the electric spark, failed to show anything of the kind when applied to the solar spectrum ; " and then he goes on to say that, in making observations by means of the electric spark, he had found that in the case of a spark taken between the poles of an induction coil, or between the poles of an electric lamp, that the visible spectrum which was revealed and rendered visible by means of fluorescence was no less than six or eight times longer than the whole of the visible part of the spectrum. This was a revelation of the first order. He was so astonished at it, that he at first thought there was some mistake. " I could not help suspecting that it was a mistake, arising from the reflection of stray light." In fact, so astonished was he, so many methods did he try in order to break down the impossibility, if it existed, that he adds, in a subsequent part of the paper, " I tried different methods,

* On the long spectrum of the electric light. *Phil. Trans.* vol. clii. p. 599.

without being able to satisfy myself as to the accuracy
of the observations, and frequently thought of resorting
to photography."

§ 4. *Dr. Miller's Work.*

While Prof. Stokes was thinking of resorting to
photography, Dr. Miller, of King's College, was not
only thinking of it but had actually resorted to it, and
was taking photographs of the so-called invisible part
of the spectrum, in which the spectrum in the case of
some substances was five or six times, and in the case
of silver one might say almost seven times, as long as
the spectrum ordinarily visible through glass prisms.
Prof. Miller went very nearly over the same ground
that Prof. Stokes had done before him.   He investigated the transparency of quartz, and came to the
conclusion that quartz is almost the only substance
that can be employed.   He also * gives for the first
time a detailed account of the way in which such work
is done.   We have first a spark from an induction coil,
between poles composed of the metals the spectra of
which Dr. Miller wished to examine.   A quartz lens to
throw the image of these poles on to the photographic
plate, prisms of quartz, and a camera ; so that he had,
first of all, a light source by which he got an intense
illumination, then a quartz train as it is called, and then
simply the photographic plate.   Having therefore an
entire absence of non-transparent glass, Prof. Miller was

* Vol. cit. p. 801.

delighted to find that, on taking the spark in this way,
between electrodes of different substances, he not only
photographed part of what could be seen, namely, a
spectrum ranging from green to blue, but one extending
as a rule six times the length of the visible spectrum
beyond the blue; although, in some cases, it is true it
was only four times as long on the more refrangible side
of H, as H is from the red end of the spectrum, that
is to say from the line A.

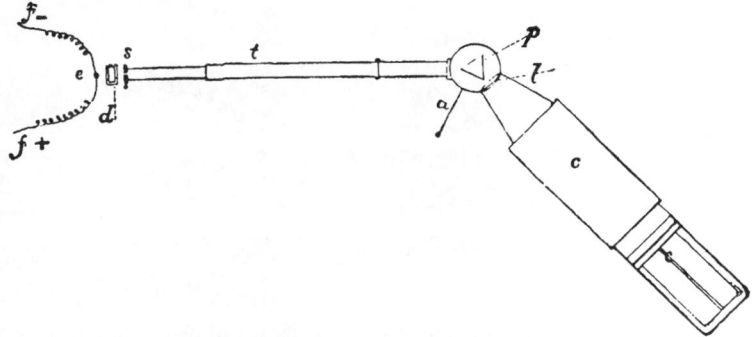

FIG. 30.—Dr. Miller's arrangements,—*s*, slit ; *l*, quartz lens ; *c*, camera ; *p*. quartz prism
t, collimator ; *e*. spark.

In this paper of Dr. Miller's we have the germ of all
the applications of photography to spectroscopic inquiry which have been carried on since; and I am
sorry to say that altogether too little has been carried
on. Not only did Dr. Miller investigate in this way the
radiation of different vapours, and give photographs for
the first time of the bright lines of a very large number
of chemical elements, but he went further than this, and
dealt with the absorption of many substances.

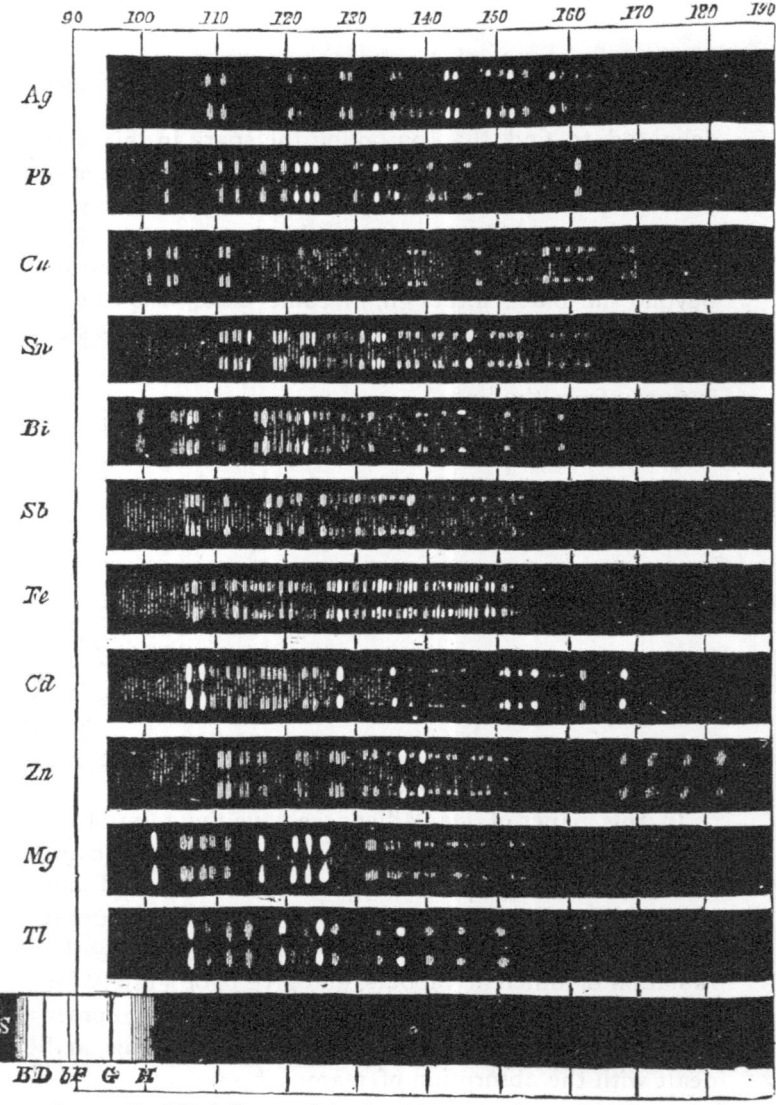

FIG. 31.—Dr. Miller's diagrams of the ultra-violet spectra of some of the elements in air.

First, he investigated the absorption of the chemical rays by transmission through different media ; through solids (transparent, of course), through liquids, and through gases and vapours; the only alteration he made in his general mode of experimentation being

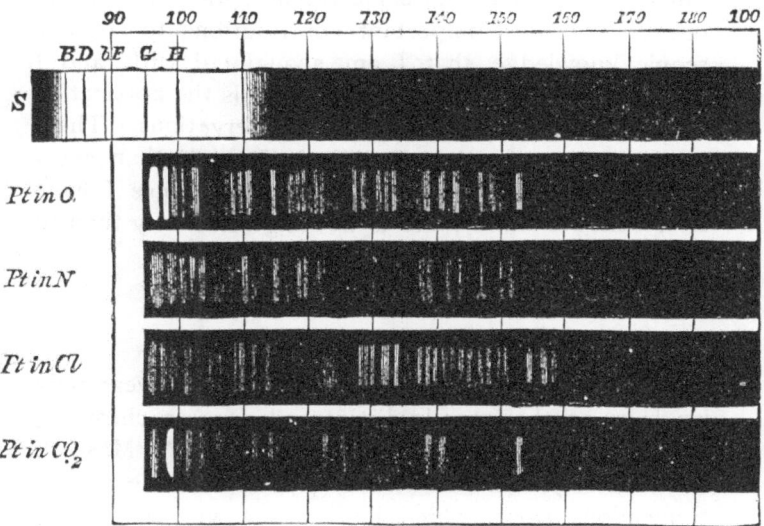

FIG. 32*.—Copies of Dr. Miller's maps of the ultra-violet spectrum of platinum in various gases showing the length of the solar and ultra-violet spectrum.

that in the case of the absorption of gases and vapours he placed the instrument further from the light source, and in the path of the ray inserted a tube containing the gas or vapour to be experimented with, so that the

* These maps have been obligingly placed at my disposal by Messrs. Longmans. J. N. L.

light which passed to the telescope was compelled to
traverse a thickness of vapour according to the length
of the tube employed.   In that way he not only determined the absorption of equal lengths of different
vapours amongst themselves, but the absorption of
different lengths of the same vapour ; his paper is thus
one of the most important contributions to spectroscopic knowledge that I am acquainted with, and I
hold that the chief importance of it is the application
of photography to spectroscopic observations.   There
are few observations more difficult, I think, than spectroscopic ones, while from the little experience I have
had, I should think there is nothing more easy than to
produce passable photographs of spectra.

§ 5 *M. Mascart's Maps.*

That, then, was in the year 1862.   In the year 1863
we have another equally distinct advance to chronicle,
but this time the work is done in France.   M. Mascart,
a name very well known to physicists—undertook a
tremendous work, which he has not yet completed,
namely, a complete investigation of the ultra violet
solar spectrum.*   Instead of using a quartz prism, as
Dr. Miller had done before him, M. Mascart uses a
diffraction grating, that is to say an instrument by
means of which the light is not refracted, as in the case
of the prism, but diffracted by an effect of interference of

* "Annales scientifiques de l'Ecole normale Supérieure."   Vol. for 1864,
p. 219.

fine lines ruled on glass. M. Mascart has shown it to be possible, by reflecting light from the first surface of the diffraction gratings, to get light diffracted without its going through glass at all. In this way, therefore, we avoid altogether the imperfect transparency of glass. Prof. Mascart has gone on advancing every year, until now he has completed, almost entirely·by means of photography, a map of the solar spectrum extending about as far as the line called T. There he finds the solar spectrum ends ; but in the case of a great many vapours, such, for instance, as that of cadmium and others of like nature, he finds he can go on photographing very much further, and he has been able to photograph to a distance, five or six, or even seven times as far from the line H as H is from A.

§ 6. *Rutherfurd's Photograph of the Solar Spectrum.*

A very beautiful reflex action of spectroscopy on photography may now be referred to for a moment. Rutherfurd, whose name is associated with that of Delarue in the matter of celestial photography, was not pleased with the action of reflecting telescopes. He lives in New York, and I suppose New York is as bad as London for tarnishing everything that the smoke and atmosphere can get at ; and he came to the conclusion that he must abstain from celestial photography altogether, or else make a lens—and a lens with Mr. Rutherfurd means something over 12 in.

diameter—which should give him as perfect an image in New York with 15in. of glass, as a perfect reflector of equal aperture.

Mr. Rutherfurd, who never minces matters, knowing that it was absolutely impossible to get such a lens as this from an optician, who of course neglects almost entirely the violet rays—the very rays which he wanted —in constructing an ordinary telescope, determined to make such an one himself. Thinking about the matter, he came to the conclusion that in any attempt to correct a lens of this magnitude for the chemical rays, the use of the spectroscope would be invaluable. He therefore had a large spectroscope constructed, in order to make a large telescope, and as a result of this we have as distinct an improvement upon the instruments which we owe to the skill of those who first adopted the suggestion of Sir John Herschel and brought together the chemical and the visual rays, as the improvement we owe to Herschel was upon the instruments which dealt simply with the visual rays.

Mr. Rutherfurd simply discards the visual rays, and brings together the violet ones; the result of his work being a telescope through which it is impossible to see anything, but through which the minutest star, down, I believe, to the tenth magnitude, can be photographed with the most perfect sharpness. This is the instrument of the future, so far as stellar astronomy is concerned.

Having thus achieved what he wished in the construction of this instrument, and having the spectroscope, Mr. Rutherfurd commenced a research which, I am sorry to say, he has never published, for it would be of the greatest value, upon the best kinds of collodion and the best arrangement of lenses for spectrum photography. He found that some collodions are so local in their action as to be almost useless for that reason, and that others are so general in their action that they are also almost useless for the exactly opposite reason.

Mr. Rutherfurd's contribution to photographic spectroscopy, his photograph of the solar spectrum from F to H, is quite as admirable and excellent as his photograph of the moon : the photograph is a refraction photograph, that is to say prisms were used, and, more than this, the prisms were of glass. It therefore, extends only a very little distance beyond the H lines. But America was not satisfied with this, and in the person of Dr. Draper, the son of the Professor Draper whose name is so honourably associated with the commencement of work done in photography thirty years ago, has just now photographed the solar spectrum far beyond H with a Rutherfurd diffraction grating.

§ 7. *The Use of the Compound Slit.*

We have already seen how exceedingly important it was to use a slit instead of a round hole in these experiments. It was shown by Wollaston with regard to eye observation, and by Becquerel and Draper with regard to spectrum photography.

It is obvious that when we observe a spectrum its breadth will depend upon the length of the slit. If we could at the same time illuminate different portions of the slit with rays proceeding from different vapours, the spectra of the different light-sources could be seen at once. But we can only do this to a certain extent. What we can do when we introduce photography is to illuminate different portions *successively*, the effect being that on different portions of the photographic plate the various spectra will successively record themselves.

Now we can take a long slit and divide it into as many portions as we choose. All we have to do is to let a window run down it, and when the window has arrived at the second part of the slit, let in light from a new source, and so on for the remaining portions.

This principle has been carried out practically in the following manner :—A rectangular brass plate 71 mm. long, and 35 mm. broad, slides in grooves in front of the slit of the spectroscope, and a window 4 mm. high, cut out of this plate, leaves a portion of the slit of this length exposed. A small pin presses firmly against the face of one of the sliding plates, and a row of small shallow holes or notches is drilled in the plate so as to intercept it in its upward or downward movement at those points where the pin falls into a notch. The distance between the notches is precisely the same as the height of the opening cut in the sliding plate, so that the movement of the plate from

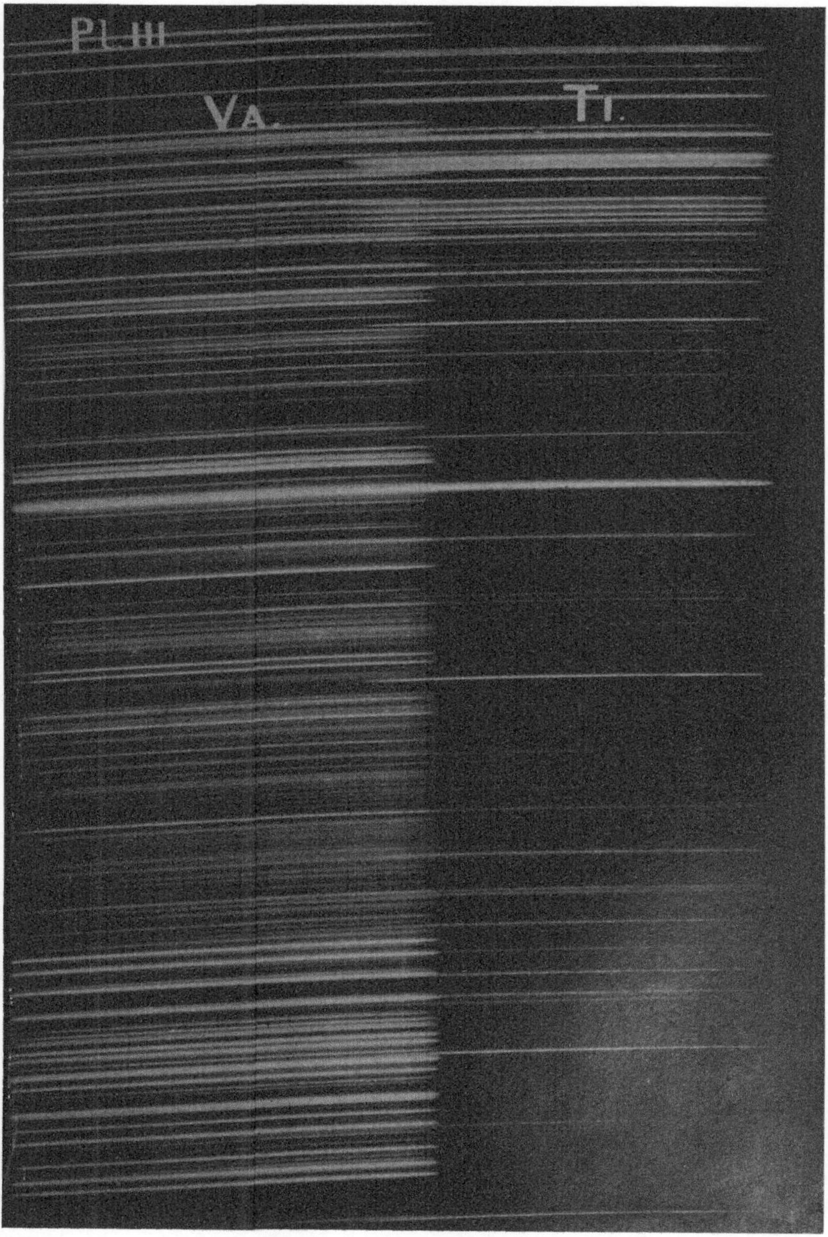

Pl. III. Va. Ti.

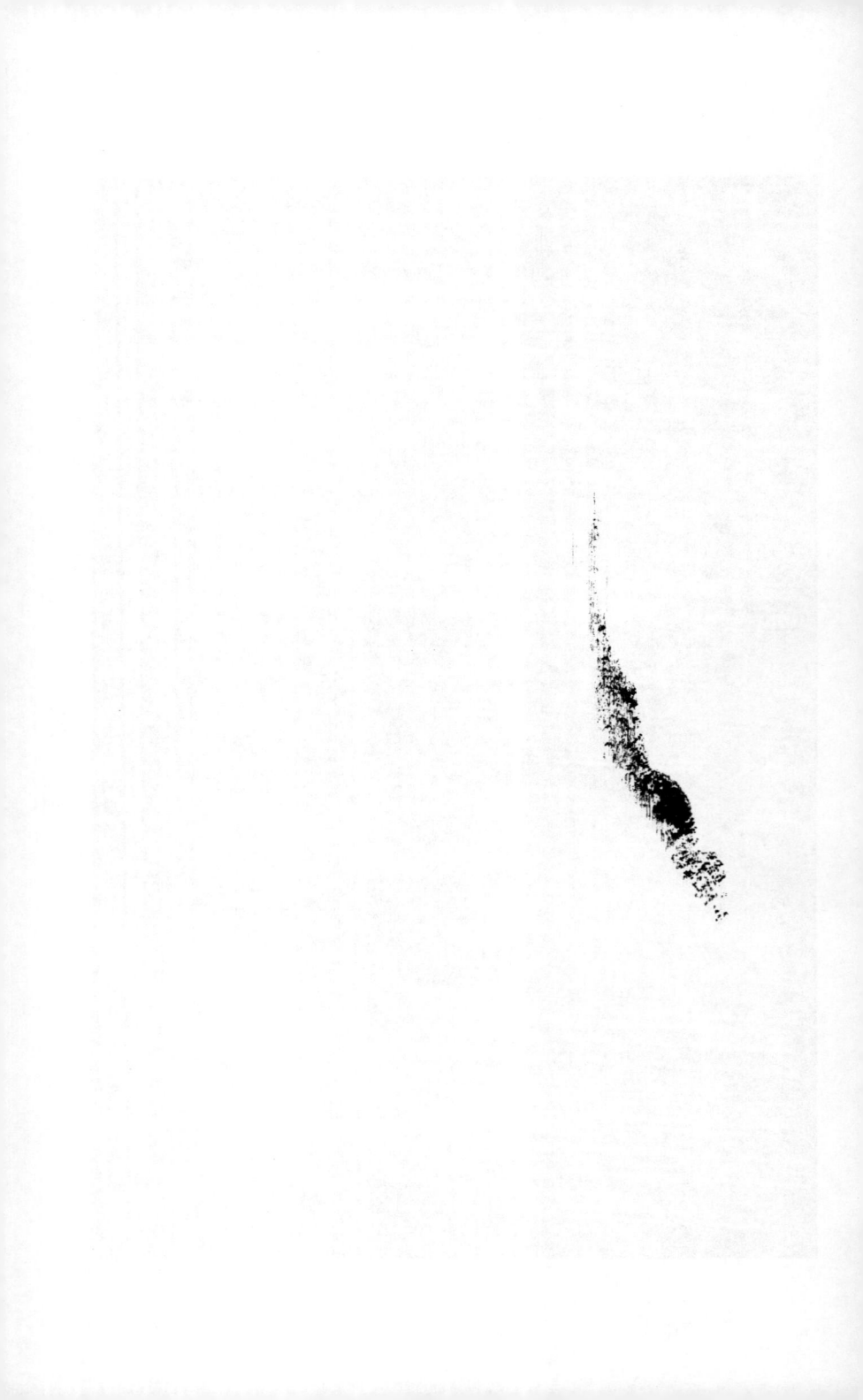

one notch to another corresponds to a distance equal to
the height of the exposed part of the slit, and the spectra compared are confronted, so to speak, absolutely ;
the upper edge of one spectrum abuts against the lower
edge of the other, and the coincidence, or want of coin-

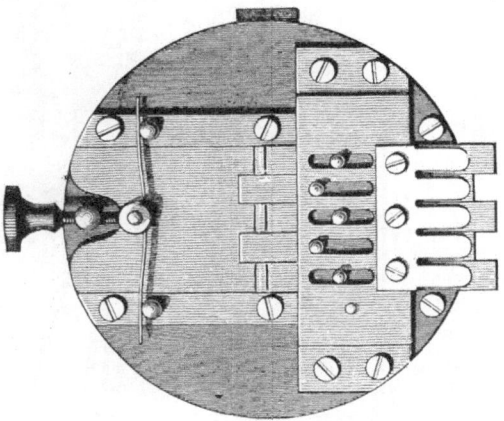

F IG. 33.—Slit allowing five photographs to be taken on the same plate by using successively different portions.   Two portions of the slit are shown covered.

cidence, between lines in the two spectra can thus be
determined with the greatest precision.   See Plate III.

Another arrangement is one shown in Fig. 33, by
which different portions of the slit are exposed by
moving exterior shutters laterally.

Having a slit of a certain length, if I open all the
length of that slit at once I should get a spectrum the
breadth of which would depend upon the length of the

H

FIG. 34.—General view of spectrum photographic arrangements, showing heliostat lamp and lenses.

slit; but if I commence operations by allowing the light to come first through one small portion of the slit, then we shall get the light from the particular metal which I employ in the electric arc falling on one part of the plate, and registering itself on the photographic plate. Then, if I close up that part of the slit, and open another one, I shall be able, through that newly opened part of the slit, all the rest being closed, to photograph on the plate the spectrum of another substance ; say iron. Then, having used up that part of the plate, I can close that portion of the slit, I can bring my window lower down, and there obtain, say the spectrum of cobalt. The window is next brought farther down, for —say—the spectrum of nickel, so that we have, as the work of some eight or nine minutes at the outside, a photograph which will register with the most absolute and complete accuracy and certainty hundreds of lines. Now a careful student of those lines, working as hard as he can, thinks himself very fortunate if he can lay down ten an hour. Therefore, as ten in an hour is to hundreds in seven minutes, so is the eye to photography in these matters.

§ 8. *My Arrangements.*

The spectroscope employed contains three prisms of 45° and one of 60°; its observing telescope is replaced by a camera with a quartz lens of 2-in. aperture, of about 5-feet focal length. With this arrangement—the spectrum being received upon a sensitized $\frac{1}{4}$ plate—the por-

H 2

tion between the wave-lengths 39,00 and 45,00, can be obtained at once in good focus. A ray of sunlight, reflected from a heliostat mirror so as to fall upon the slit-plate, is brought to a focus, by means of a double convex lens, just between the carbon poles of an electric lamp, while a second convex lens, placed between the lamp and the collimator tube, serves to cast an image of the sun or of the electric arc upon the slit-plate. Supposing, now, we wish to compare the iron spectrum with that of the sun : the sun's image in sharp focus on the slit-plate is first allowed to imprint its spectrum on the prepared plate. The ray of sunlight is then cut off, the sliding plate moved up or down till the pin catches in the next notch, and the image of the arc, between an upper pole of carbon and a lower pole consisting of a carbon crucible containing a fragment of iron, is allowed to fall on the portion of the slit thus exposed.

The following details will render the application of the method quite clear :—The laboratory in which the work has been carried on has two windows, one nearly (magnetic) south, the other nearly (magnetic) west. Outside each window level slate slabs have been erected as supports for a heliostat. Either window can be used at pleasure. The spectroscope is supported on a platform on rollers, the height of the platform being such that the horizontal beam from the heliostat is coincident with the axis of the collimator. In addition to the lens placed between the lamp and the slit to throw an image of the arc on the latter, another lens is introduced

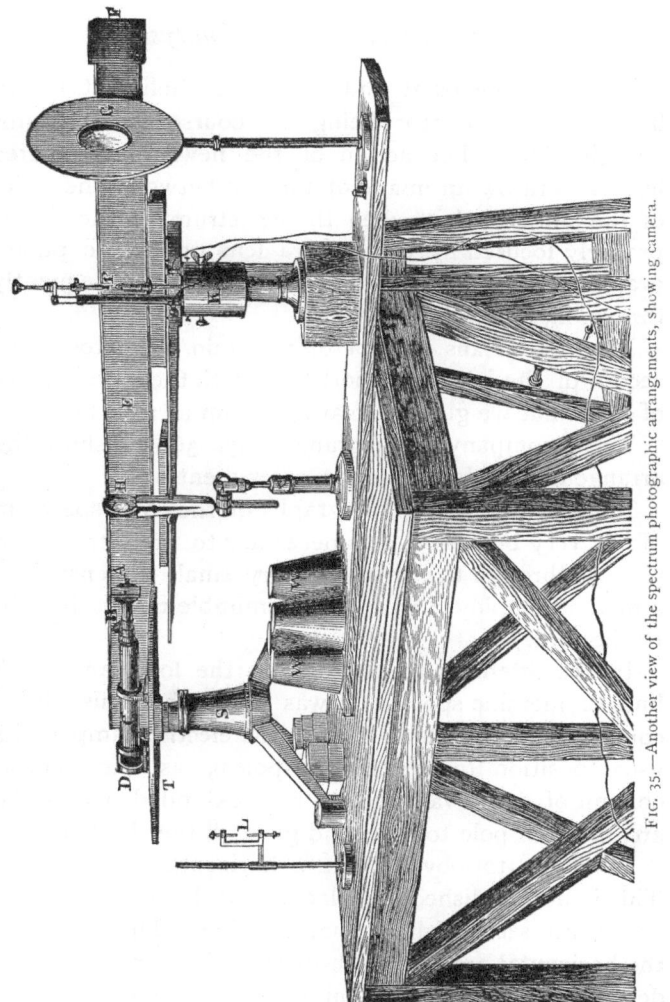

FIG. 35.—Another view of the spectrum photographic arrangements, showing camera.

between the heliostat and the lamp—heliostat, lenses, lamp, and collimator being of course in the same straight line.   The action of the newly interpolated lens is to throw an image of the sun between the poles of the lamp, so that when the spectrum of the arc is properly focused by the camera lens on to the photographic plate, the solar spectrum, when subsequently thrown in, is also in focus.

By these means we not only obtain a photographic record of the long and short lines with the individuality of each, but we get the solar spectrum as a scale.

The accompanying diagrams (Figs. 36–38) show the arrangements adopted in the cases mentioned.

In order to obtain photographs of the solar spectrum of the very best kind, it is necessary to limit the beam passing through the prisms to very small dimensions— a method employed with such admirable results by Mr. Rutherfurd.

In the attempts to photograph the long and short lines of metallic spectra, it was found that this object could not be well obtained with the electric lamp in its usual position (with vertical poles), as the central column of dense vapour, as a rule, extended across the arc, *i.e.* from pole to pole, and gave all the short lines.

In order to obviate this a horizontal arc is used. This is accomplished by placing the lamp on its side and firmly securing it in that position.   The image of the horizontal arc is then thrown on the vertical slit as described in Chapter II.   This method is found to be perfectly successful.   The central portion of the spectrum

Fig. 36.—Arrangement for obtaining solar spectrum alone.

Fig. 37.—Arrangement for obtaining long and short lines.

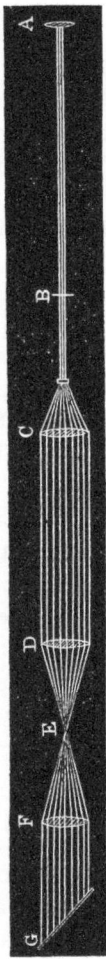

Fig. 38.—Arrangement for obtaining and comparing lines with solar spectrum.—A, collimating lens; B, slit; C, opera-glass; G, heliostat; D, lens; E, poles; F, lens throwing image of sun between poles.

due to the dense core of the arc is found to consist of all the lines which this part of the arc alone can give, the longer lines given by the outer less dense strata extending beyond the spectrum of the core to the various distances from the centre to which the vapour capable of giving them extends.

The lines thus photographed are pointed at either end, and disappear from the centre in the order of their lengths, so that a double or duplicate determination, exquisitely symmetrical, of length is thus obtained.

Although the lengths, thicknesses, and intensities of the lines are thus readily recorded, we have so far no scale by which to fix their positions. In order to obviate this objection, the solar spectrum is photographed on each plate immediately above or below the metallic spectrum under examination.

This new method has enabled me to register no less than four different spectra on the same wet plate. Those familiar with photographic processes will immediately see how it is that the number is not forty instead of four.

§ 9. *Results of these Arrangements.*

We have thus an absolute comparison rendered possible, by means of photography, between the lines of the spectrum of a metal, and the lines of the spectrum of the sun. Plate IV. gives an idea of such photographs. In the case of most of the thick lines we get a thick line in the solar spectrum corresponding with the

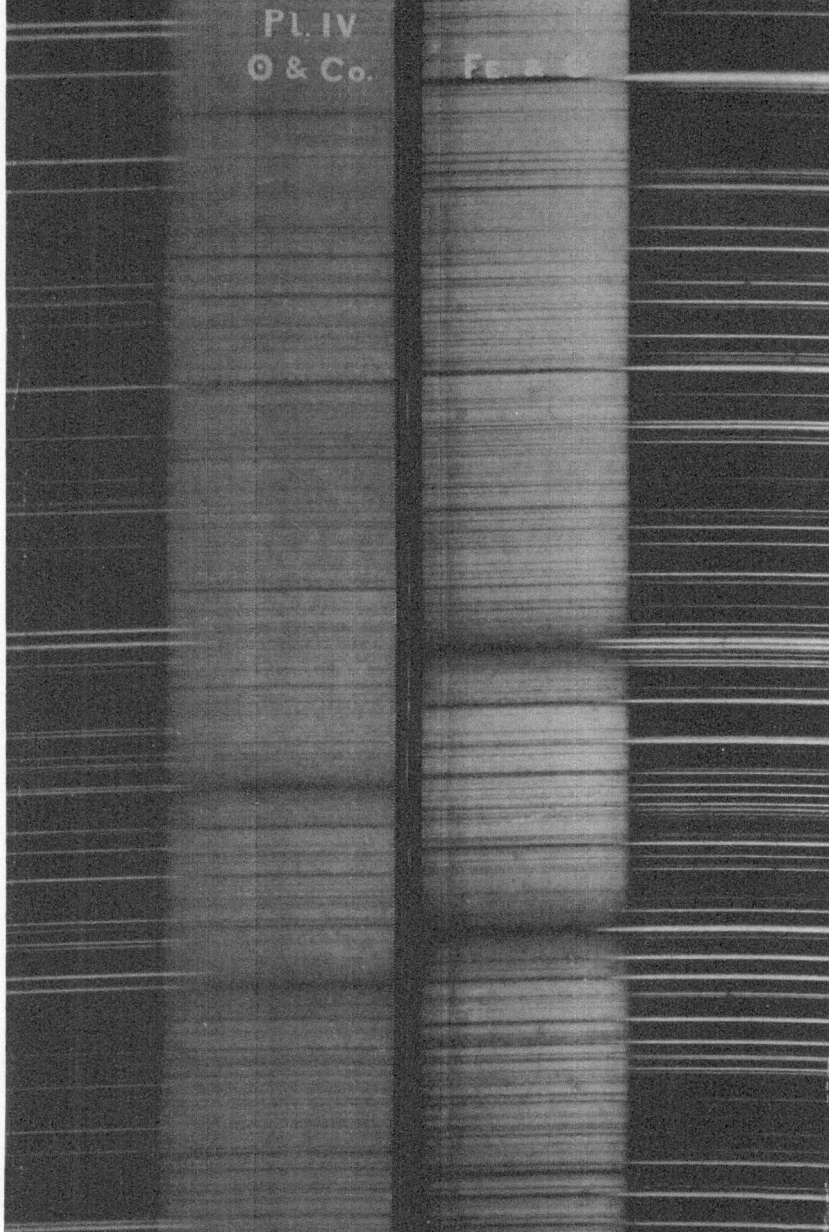

Pl. IV
O & Co.   Fe & Co.

lines of the metal. In some cases the lines of the metal are of different lengths. The reason of that is, that care has been taken to photograph on the plate the lines due to the various strata of the vapour, from the rarest vapour, which is obtained at the outside of the electric arc, to the densest, which occupies the centre of the core. We thus get a most beautiful gradation, as we pass from the outside part of the spectrum to the inside. The inside part represents the complete spectrum of the core, and the outside the incomplete and almost monochromatic spectrum of the vapour which surrounds the denser core in the middle of the spark ; thus we practically reduce the spectrum of the metal to one line.

§ 10. *The Solar Spectrum.*

We may next consider the application of spectrum photography no longer to the mere solar spectrum, but to the physics of the sun. What is the solar spectrum ? It is the continuous spectrum of the sun, minus certain portions where the light of the continuous spectrum has been absorbed. What have been the absorbers ? The gases and vapours, generally speaking, in an excessively limited zone of the sun's atmosphere, lying close to the bright sun we see—that is close to the photosphere. This zone is called the reversing layer. Then if the solar spectrum is the result of the absorption of this reversing layer, what will happen to the solar spectrum if the constitution of the layer changes ? Obviously a

change in the solar spectrum. Now, the researches referred to in Chapter VIII., researches carried on by means
of photography—show us that if we take any particular vapour in the reversing layer, which we may call A,
for instance, and then assume that the quantity of A in
the layer is reduced, the absorption of that particular
vapour will be reduced ; what then will be the result
on the photograph of the solar spectrum ? Some of
the lines will disappear. Suppose that this particular
vapour which we called A, instead of being assumed to
decrease in quantity, increases in quantity, what will
happen to the solar spectrum ? The same researches
tell us that as its quantity increases its absorption will
increase, and that its increased absorption will be indicated by an increase in the number and in the breadth
of the lines absorbed. What, then, will happen to the
solar spectrum if any change of this kind is going on ?
The photograph of a solar spectrum taken, say, to-day,
may be different from the photograph of the same part
of the spectrum taken at some distant period. What is
the distant period we do not yet know—whether three
months, six months, six years, or eleven years, or a multiple of eleven years—but, at all events, there is reason
to think already that if we had a series of photographs
of the solar spectrum taken year by year, we should
see changes in the spectrum. A photograph of a
very limited portion of the solar spectrum will prove
my case ; and I could not have proved it if photography
had not been called in, because if the existence of any
particular metal, or of the increase of any particular

metal, depends upon such a small matter as one line among 10,000, what will happen if a man neglects to observe this change ? People will say, " Oh ! in a research of that kind it is altogether excusable if he has made a mistake." But if we have a series of phenomena recorded by means of a camera on " a retina which never forgets," as Mr. Delarue has beautifully put it, and if we compare those pictures day by day and year by year, the thing is put beyond all question when we observe one line disappearing, or another line appearing.

I have already given a drawing of the part of the solar spectrum near the line H. I wish to call attention to one line in this region. We have an admirable map of the solar spectrum made about the year 1860. The draughtsman, recording by means of his eye the lines in the spectrum, would not be very likely to overlook a line darker than some he inserts, but he might easily overlook finer lines. Now, it is a fact that in the most careful map that we have—a map drawn with a most wonderful honesty and splendid skill—a line is absent in the region indicated, which line is now darker than some that were then drawn, and that line indicates the presence of an additional element in the sun—strontium.

I do not make this assertion thinking that subsequent facts will show the drawing to be wrong, but because I see reason to believe that what we know already of the sun teaches us that it is one of the most likely things in the world that strontium was not present in such great quantity in the reversing layer when the drawing was made ; but, however that may be, all must recognize how important it is that photographs should be compared with photographs made five, ten, fifteen, a hundred, or two hundred, or as many years as you like apart, and it is in this possible continuity of observation of the solar spectrum, carried on for centuries, that I think we have in photography not a tremendous ally of the spectroscope, but a part of the spectroscope itself. Spectroscopy, I think, has already arrived at such a point, at all events in connection with the heavenly bodies, that it is almost useless unless the record is a photographic one.

§ 11. *Stellar Photography.*

Dr. Draper and Dr. Huggins have succeeded in getting photographs of the spectra of some of the stars. This is a matter of the very highest importance, because the sun is nothing but a star, and the stars are nothing but distant suns ; and as long as we merely investigate our sun, however diligently or admirably we do it, and neglect all the others, it is as if a man who might have the whole realm of literature to work at should confine himself to one book, and that book probably not a good representative of the literature of the country he was inquiring into.

To the spectroscope all nature is one, and it is absolutely impossible to make a single observation, either on a sun, or a star, or a comet, without bringing chemical and

physical considerations into play ; and it will be a re-
grettable circumstance if chemists employ the spectroscope in terrestrial chemistry—they have not done much
in that way yet—without taking the sun and all the
various stars of heaven into counsel, because the spectroscope is absolutely regardless of space, and tells us
that the elements which are most familiar to us here, or
at all events a good many of them, are present in the
most distant stars, and the spectroscope shows us those
elements existing under conditions which further are
absolutely impossible here.

§ 12. *The Work of the Future.*

We have, I believe, what we may almost call a new
chemistry, some day to be revealed to us by means of
photographic records of the behaviour of molecules.
Recollect that the difference between the iron spectrum
of one line and the iron spectrum of between 400 and
500 lines is simply due to the difference in the vibration of the molecules or atoms of iron in the centre
of the electric arc and its exterior. There is one question which all students of the spectroscope may ask of
photographers, and it is this. Why should we any
longer be confined, in registering spectra, to the more
refrangible end of the spectrum, when one of the very
first spectra of the sun that was ever taken was a complete photograph of the spectrum, including not only
the blue and the green, but the yellow, the red, and the
extreme red ? I think that if photographers will study

the action of light on molecules, and will give those who are familiar with the spectroscope, and those who are anxious to promote the progress of spectroscopic research, a means of extending photographic registration, not only into the green part of the spectrum which they do already with difficulty, but to the extreme red, then the use of the eye will almost entirely be abolished in these inquiries. And although no one has a higher estimate than myself of the extreme importance of the eye, I think that the more it is replaced by permanent natural records in these inquiries, the better it will be for the progress of Science.

As I have already hinted, I believe the spectroscope is capable of leading along the track of discovery in this matter. It has begun, I think, by altogether abolishing the distinction drawn in all our text-books between the chemical and the visible rays. The curves which are given in these books, showing us the maxima of heat, light, and chemical action, are, I fancy, merely curves showing us, as it were, the absorption spectra of those substances by which the maxima have been determined—whether they be lamp-black, the coating of the retina, or salts of silver, and are really altogether independent of the nature of light. The salts of silver ordinarily employed require short waves to set them vibrating first, and dissociating afterwards as a result of the vibration. Why should not other salts shiver even under the influence of the ultra red? What it is that renders a molecule apt to vibrate with one wave-length more than another remains to be discovered; there is

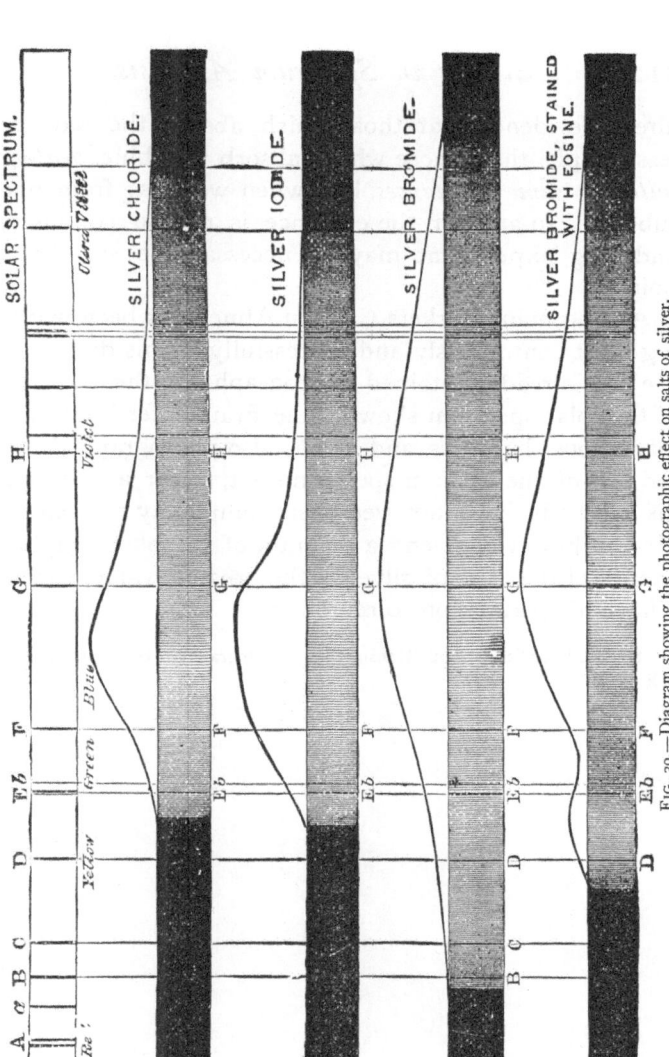

FIG. 39.—Diagram showing the photographic effect on salts of silver.

already evidence that those which absorb the red are less simple than those which absorb the blue, *dealing with the same substances*, but when we pass from one substance to another the evidence is not so complete, and long experiment may be necessary to settle the point.

Among many workers Captain Abney has been working most continuously and successfully in this direction. He has already obtained photographs of the red end of the solar spectrum showing the Fraunhofer lines with marvellous sharpness, and he has also photographed the red end of the calcium spectrum, but, so far as I know, his methods have not yet been completely published. He has, however, given* a diagram of the photographic energy of the salts of silver of the greatest value, which I have his permission to reproduce.

* Science Lectures at South Kensington. Photography. By Capt. Abney, R.E., F.R.S.

CHAPTER IV.

ATOMS AND MOLECULES SPECTROSCOPICALLY CONSIDERED.

WHAT are atoms and what are molecules? A chemist discourses on the atomic weight of certain elements, and defines and talks about molecular volumes, and the like. Here is a definition given by Dr. Frankland in his book on Chemistry ("Lecture Notes," p. 2): "An atom is the smallest proportion by weight in which the element (that is to say the element to which the atom under discussion belongs) enters into or is expelled from a chemical compound." He then points out that when atoms are isolated—that is, when they are separated from other kinds of matter — they do not necessarily exist as atoms in the old sense; they go about in company, generally being associated in pairs. He then defines such a combination of atoms as an elementary molecule. Here, then, is put before us authoritatively a chemist's view of the difference between an atom and a molecule.

Let us now go to the physicist and see if we can gather from him his idea of atoms or molecules. It is remarkable that, in Prof. Clerk-Maxwell's " Theory of Heat," in which we find much that is known by

I

physicists about molecular theories, the word "atom" is not used at all. We are at once introduced to the word "molecule," which is defined to be "a small mass of matter the parts of which do not part company during the excursions which the molecule makes when the body to which it belongs is hot."

Prof. Clerk-Maxwell goes on to give us ideas about these "molecules," which have resulted from the investigations of himself and others. Here are some of them (p. 286): "All bodies consist of a number of small parts called molecules. Every molecule consists of a definite quantity of matter, which is exactly the same for all the molecules of the same substance. The mode in which the molecule is bound together is the same for all molecules of the same substance. A molecule may consist of several distinct portions of matter held together by chemical bonds, and may be set in vibration, rotation, or any other kind of relative motion, but so long as the different portions do not part company but travel together in the excursions made by the molecule, our theory calls the whole connected mass a single molecule." Here, then, we have our definition of a molecule enlarged.

The next point insisted upon is that the molecules of all bodies are in a state of continual agitation.

That this agitation or motion exists in the smallest parts of bodies is partly made clear by the fact that we cannot see the bodies themselves move.

What, then, on this theory, is the difference between the solid, liquid, and gaseous states of matter?

In a solid body the molecule never gets beyond a certain distance from its initial position. The path it describes is often within a very small region of space. Prof. Clifford, in a lecture upon atoms, has illustrated this very clearly. He supposes a body in the middle of the room held by elastic bands to the ceiling and the floor, and in the same manner to each side of the room. Now pull the body from its place ; it will vibrate, but always about a mean position ; it will not travel bodily out of its place. It will always go back again.

We next come to fluids : concerning these we read— " In fluids, on the other hand, there is no such restriction to the excursions of a molecule. It is true that the molecule generally can travel but a very small distance before its path is disturbed by an encounter with some other molecule ; but after this encounter there is nothing which determines the molecule rather to return towards the place from whence it came than to push its way into new regions. Hence in fluids the path of a molecule is not confined within a limited region, as in the case of solids, but may penetrate to any part of the space occupied by the fluid."

Now we have the motion of the molecule in the solid and the fluid. How about the movement in a gas ? " A gaseous body is supposed to consist of a large number of molecules moving very rapidly." For instance, the molecules of air travel about twenty miles in a minute. " During the greater part of their course these molecules are not acted upon by any sensible force, and therefore move in straight lines with uniform velocity. When two molecules come within a certain distance of each other, a mutual action takes place between them which may be compared to the collision of two billiard balls. Each molecule has its course changed and starts in a new path."

The collision between two molecules is defined as an "Encounter;" the course of a molecule between encounters a "Free path." It is then pointed out that "in ordinary gases the free motion of a molecule takes up much more time than is occupied by an encounter. As the density of the gas increases the free path diminishes, and in liquids no part of the course of a molecule can be spoken of as its free path."

The Kinetic Theory of Gases, on which theory these statements are made, has this great advantage about it, that it explains certain facts which had been got at experimentally, facts which had been established over and over again, but which lacked explanation altogether till this molecular theory, which takes for granted the existence of certain small things which are moving rapidly in gases, less rapidly in fluids, and still less in solids, was launched. The theory, in fact, explains in a most ample manner, many phenomena so well known that they are termed Laws. It explains Boyle's law, and others.

This theory, which takes for its basis the existence of molecules and their motions, explains pressure by likening it to the bombardment of the sides of the containing vessel by the molecules in motion; or it tells us that the temperature of a gas depends upon the velocity of the agitation of the molecules, and that this velocity of the molecules in the same gas is the same for the same temperature, whatever be the density. When the density varies, the pressure varies in the same proportion. This is Boyle's law. Further, the densities of two gases at the same temperature and pressure are proportional to the masses of their individual molecules, or, when two gases are at the same pressure and temperature, the number of molecules in unit of volume is the same. This is the law of Gay Lussac.

We are now, then, fairly introduced to the "atom" of the chemist and the "molecule" of the physicist ; we see at once that the methods of study employed by chemical and physical investigators are widely different. The chemist never thinks about encounters, and the physicist is careless as to atomic weight ; in his mind's eye he sees a perpetual clashing and rushing of particles of matter, and he deals rather with the quality of the various motions than with the material.

In Prof. Clerk-Maxwell's book (p. 306) it is assumed that while the molecule is traversing its free path after an encounter, it vibrates according to its own law, the law being determined by the construction of the molecule, or let us say its chemical nature, so that the vibration of one particle of sodium would be like that of another particle of sodium, but unlike that of a particle of another chemical substance, let us say iron. If the interval between encounters is long, the molecule may have used up its vibrations before the second encounter, and may not vibrate at all for a certain time previous to it. The extent of the vibration will depend upon the kind of encounter, and will in a certain sense be independent of the number of encounters.

We can imagine a small number of feeble encounters, a large number of feeble encounters, a small number of strong encounters, and a large number of strong encounters.

In the case of feeble encounters, we pass from a small number to a large one by increasing the density.

In the case of strong encounters we pass from low temperature with small density to high temperature with great density.

Increase of density will reduce " free path."

Increase of temperature will increase the intensity of the vibrations.

The shorter the free path the more complex the vibrations.

The more decided the encounter the more will the vibration of the molecule be brought out, not merely the *fundamental vibrations*, as we may term them, which we get when the free path is longest, but all those which are possible to each molecule.

Now why have these detailed statements concerning the vibrations of molecules been necessary? Because we believe that each molecular vibration disturbs the ether; that spectra are thus begotten; each wave-length of light resulting from a molecular tremor of corresponding wave-length. The molecule is, in fact, the sender, the ether the wire, and the eye the receiving instrument, in this new telegraphy.

As before, let us endeavour to see if our ideas may be rendered more clear by any of the more familiar phenomena in sound.

Let us call the motions, of whatever nature they may be, set up in a molecule of matter, vibrations. Then, to get the most concrete notion of such a light source, let us compare it with the most simple sound source, a tuning-fork.

The same tuning-fork will always give us the same sound, but the sound is more complex than might at first sight be imagined. In addition to the prevailing note, which depends upon the number of vibrations per second, as we have seen, there are other tones which have a definite relationship to the prevailing, or, as it is termed, the fundamental note. The loudness of all these, as we have also seen, depends upon the amplitude of the vibration of the tuning-fork.

But there are more complicated cases of vibration than these.

In the violin we have a convenient example which shows us that the quality of the sound produced, that is the quality of the vibration of the string set up, depends upon the manner in which the bow is drawn over the string.

Similarly, the same vibrating plate, when damped at different points, vibrates in quite a different manner.

Now does the spectroscope throw any light upon molecular questions? is there any hope that the spectroscope, as researches with it are extended, may aid the study of a subject which lies at the root of chemical and physical investigation ?

In order to endeavour to answer these questions, I will now proceed to lay down some propositions embracing the knowledge which has been acquired up to this time from this point of view. These propositions I shall take one by one. I shall state the experimental basis on which the statements rest, and shall refer to the methods by which the results have been obtained.

I shall for a time use the word " particle " to represent a small mass of matter, because it does not tie me to the " atom," or the " molecule " of the chemist, or to the "molecule" of the physicist. "Particle" is a neutral term, which I hope none will quarrel with.

§ 1. *Radiation.*

1. *When bodies retain a solid or liquid form when incandescent, their constituent molecules give out rays of light such that the spectrum is continuous as far as it goes.*

This was Kirchoff's first generalization.

It surely is an important fact from the point of view of the molecular theory, that all solids and liquids, with their particles moving in the manner already stated, do give us a perfectly distinct spectrum from that which we get when we deal with any rare gas or vapour whatever. A poker put into the fire becomes first of a dull red heat, after a time a white heat is arrived at. So far as the vibrations exist they are continuous, there are no breaks in the series of wave-lengths. We may also drive a platinum wire to incandescence in the same way by means of electricity. Analyse the light by means of the spectroscope, the spectrum is the same as that of the poker. Further, we can go to the sun, and divest it in imagination of the atmosphere which absorbs so much of its light, and we know that, with a small exception, we shall get a perfectly continuous spectrum similar to that in the case of the poker or platinum wire.

In this continuous spectrum we have a spectroscopic fact connected with that kind of molecular motion which physicists attribute to particles so long as they are closely packed together in the solid state, and so long as they have but a small free path, as in the fluid state.

2. *When particles are in a state of gas or vapour, and are rendered incandescent by high tension electricity, line-spectra are produced in the case of all the chemical elements.*

These line-spectra are only to be obtained from gases and vapours, and, with few exceptions, only when we employ high-tension electricity.

We get a spectroscopic result perfectly distinct from the one we had before, precisely in the case where according to the physicists we have an enormous motion and agitation of particles.

3. *The characteristic vibration of a particle is independent of length of free path.*

In the Kinetic theory, as generally enunciated, there is nothing to show that the same particle may not be in question in the solid, liquid, and gaseous states, the only change of condition being in the amount of free path.

Now as the spectra of solids and gases present a complete difference in kind (see 1 and 2), if the particle were always the same, it would be necessary to assume that, under different conditions of free path, the same particle can be thrown into different states of vibration and give us different spectra.

The question is, have we at the present time any facts at our disposal? I think we have, although some may not consider them sufficient in number or cogency; but it must always be remembered that it is precisely in such questions as these that experiments on an extended scale become almost impossible.

All the facts we have, however, tend to show that a known change of molecular condition is always accompanied by a change of spectrum, *e.g.*, sulphur vapour above and below 1,000° C.—at which point its vapour density changes—has distinct spectra.

Salts have spectra of their own, in which no lines, either of the constituent metal or metalloid, are to be found.

Another line of argument. In some cases we can mix vapours with liquids and the spectrum of the vapour remains unchanged in character; that is, the circumambient particles of the liquid behave in one case in exactly the same manner as the circumambient particles of the air do in another.

Iodine in bisulphide of carbon, $N_2 O_4$ in water, and didymium salts in water are illustrations.

Nay, we may even enclose, or appear to enclose, some substances in glass (salts of didymium, erbium, nickel, cobalt), and we get a spectrum so special in each case that we know that the particles are still going through their motions, are still vibrating, in spite of the absence of free path, and in spite of the "solid" state of their surroundings.

I shall, elsewhere, use other lines of argument to show that the reason that we so rarely see these characteristic spectra in connection with the solid state lies in the fact that the solid state is one reached not only by reduction of free path, which enables the molecules to lie nearer together, but by a reduction of molecular agitation, which in all probability enables them to combine *inter se.*

4. *In some cases particles in a state of gas or vapour can be set swinging by heat waves.*

Salts of sodium and strontium, subjected to the heat of a Bunsen burner, are at once dissociated, and the particles of the metals are set swinging by the heat waves, and we get the lines in the spectra of their vapours. Now that is not only true for salts of strontium and sodium, but for some of the elements themselves. But if salts of iron, or of the other heavy metals are placed in the flame, we do not get bright lines. Or again, in some other vapours, such as sulphur, we only get a spectrum, not of lines, but continuous over a limited part of the spectrum. In fact it may be said, with the exception of these elements which easily reverse themselves, this heat is absolutely imcompetent to give anything like a bright line.

5. *The spectra of both elementary and compound bodies vary with varying degrees of heat.*

It has already been stated that a Bunsen burner is enough to set an atom of sodium free from its combination with chlorine and make its vapour give us a bright line, while we cannot do this in the case of iron and other substances. We may say then that we have there a first stage of temperature. Many monad metals give us their line spectra at a low degree of heat. Take some dyad metals, such as zinc and cadmium ; this first stage of temperature will only make them red or white hot, a much higher temperature is required to drive them into vapour. We get the line spectrum from sodium ; do we get that from cadmium when we have melted cadmium ? We do not. This is an excessively important point. The first stage of temperature, which gives us a line spectrum in the case of sodium, is powerless to give us such a spectrum in the case of cadmium.

A second stage of heat at least is therefore required to get a line spectrum. If I take sulphur, dealing with it by means of absorption, and heat it, I get a continuous spectrum at the first stage. I increase the heat to the second stage, what do I get then ? A line spectrum, as I do in the case of sodium? No! A spectrum like that of carbon, not a line spectrum at all. I apply still a higher, a third, stage of temperature and then I get a line spectrum. In the case of the metalloids we have thus three stages of heat with three spectra. If there is such a thing as a particle at all, are we not justified in asking whether there is not some difference between the "particular" arrangements of the metalloids, and those of the metals? and some connection between temperature and the "atomic weights" of the chemist?

Before I go further I will throw these results into a tabular form, which will show that through these various heat stages, in the case of metals like sodium there is a great preponderance of line spectrum, and in the case of metalloids like sulphur, there is a great preponderance of fluted spectrum.

	Na.	Cd.	S.
Fifth stage—spark	line spectrum	line	line
Fourth stage—arc	line	line	fluted
Third stage—white heat	line	(?)	,,
Second stage—bright red heat	line	continuous absorption in the blue	,,
First stage of heat—dull red heat	line	continuous spectrum	continuous absorption in the blue

6. *From the fact that we have lines in the spectra of compound gases, it would be hazardous to affirm that the aggregate, which, with the highest dissociating power*

we can employ, gives us a line spectrum of a so-called element, could not be broken up if a still higher dissociating power could be employed.

This proposition has a bearing not only on the celestial but also on the terrestrial side of the inquiry, and is referred to at length in the chapter on dissociation.

7. *There is spectroscopic evidence which seems to show that, starting with a mass of solid elemental matter, such mass of matter is continually broken up as the temperature (including in this term the action of electricity) is raised.*

The evidence upon which I rely is furnished by the spectroscope in the region of the visible spectrum.

To begin by the extreme cases, all solids give us continuous spectra ; all vapours produced by the high-tension spark give us line-spectra.

Now the continuous spectrum may be, and as a matter of fact is, observed in the case of chemical compounds, whereas all compounds known as such are resolved by the high-tension spark into their constituent elements.　We have a right, therefore, to assume that an element in the solid state is a more complex mass than the element in a state of vapour, as its spectrum is the same as that of a mass which is known to be more complex.

The spectroscope supplies us with intermediate stages between these extremes.

(*a*) The spectra vary as we pass from the induced

current with the jar to the spark without the jar, to the voltaic arc, or to the highest temperature produced by combustion. The change is always in the same direction ; and here, again, the spectrum we obtain from elements in a state of vapour (a spectrum characterized by spaces and bands) is similar to that we obtain from vapours of which the compound nature is unquestioned.

(β) At high temperatures, produced by combustion, the vapours of some elements (which give us neither line- nor channelled space-spectra at those temperatures, although we undoubtedly get line-spectra when electricity is employed, as before stated) give us a continuous spectrum at the more refrangible end, the less refrangible end being unaffected.

(γ) At ordinary temperatures, in some cases, as in selenium, the more refrangible end is absorbed ; in others the continuous spectrum in the blue is accompanied by a continuous spectrum in the red. On the application of heat, the spectrum in the red disappears, that in the blue remains ; and further, as Faraday has shown in his researches on gold-leaf, the masses which absorb in the blue may be isolated from those which absorb in the red. It is well known that many substances known to be compounds in solution give us absorption in the blue or blue and red ; and, also, that the addition of a substance known to be compound (such as water) to substances known to be compound which absorb the blue, superadds an absorption in the red.

In those cases which do not conform to what has

been stated, the limited range of the visible spectrum must be borne in mind. Thus I have little doubt that the simple gases, at the ordinary conditions of temperature and pressure, have an absorption in the ultra-violet, and that highly compound vapours are often colourless because their absorption is beyond the red, with or without an absorption in the ultra-violet. Glass is a good case in point ; others will certainly suggest themselves as opposed to the opacity of the metals.

If we assume, in accordance with what has been stated, that the various spectra to which I have referred are really due to different molecular aggregations, we shall have the following series, going from the more simple to the more complex.

First stage of complexity of molecule . . . } Line-spectrum.

Second Stage . . . .   Fluted-spectrum.

Third stage . . . . . { Continuous absorption at the blue end not reaching to the less refrangible end. (This absorption may break up into a fluted-spectrum.)

Fourth stage . . . . . { Continuous absorption at the red end not reaching to the more refrangible end. (This absorption may break up into a fluted-spectrum.)

Fifth stage . . . . .   Unique continuous absorption.

One or two instances of the passage of spectra from one stage to another, beginning at the fifth stage, may be given.

From the fifth stage to the fourth—The absorption

of the vapours of potassium in a red-hot tube is at first continuous. As the action of the heat is continued, this continuous spectrum breaks in the middle ; one part of it retreats to the blue, the other to the red.

From the fourth stage to the third—Faraday's researches on gold-leaf best illustrate this ; but I hold that my explanation of them by masses of two degrees of complexity only is sufficient without his conclusion (" Researches in Chemistry," p. 417), that they exist " of intermediate sizes or proportions."

Gold is generally yellow, as you know, but gold is also blue and sometimes red. It must be perfectly clear to all, that if particles vibrate the colours of substances mùst have something to do with the vibrations. If the colours have anything to do with the particles it must be with their vibrations. Now as the spectrum in the main consists of red, yellow, and blue, the red and the blue rays are doing something in a substance which only transmits or reflects the yellow light ; if we put gold leaf in front of the lime light, we can see whether the yellow light does or does not suffer any change. The yellow disappears ; we have a green colour ; the red and blue are absent. The gold leaf is of excessive thickness. What would happen could I make it thinner? Its colour would become more violet. This I have proved by using aqua regia. But we can obtain a solution of fine gold, which lets the red light through. Its particles are doing something with the blue vibrations. We can obtain another solution which only transmits the blue. Now what is the difference—

K

the " particular " differences—between the gold in these solutions, and that which is yellow by reflected, and green or violet by transmitted light? It is a question worthy of much study. Here are some more experiments. Take some chloride of cobalt, which is blue, put it into a test-tube, to which add water. It turns red. I content myself by asking why it turns red? We take some chloride of nickel, which is yellow, and put it into another test-tube: we add water, and it turns green. First question—Why this change? Second question—Has the green colour of this solution anything to do with the red colour of the solution of gold?

From the third stage to the second—Sulphur-vapour first gives a continuous absorption at the blue end; on heating, this breaks up into a fluted-spectrum.

The fluted spectra of potassium and sodium make their appearance after the continuous absorption in the blue and red vanishes.

From the second stage to the first—In many metalloids the spectra, without the jar, are fluted, on throwing the jar into the circuit the line-spectrum is produced, while the cooler exterior vapour gives a fluted absorption-spectrum.

The fluted spectra of potassium and sodium change into the line-spectrum (with thick lines which thin subsequently) as the heat is continued.

These various molecular combinations may go far to explain the law of multiple proportions of the chemist.

8. *Line spectra become more complicated with increased density or temperature, provided the state of gas or vapour be retained.*

The importance of this observed fact in connection with the molecular theory cannot be overrated. In the solid the particles can only oscillate round their mean position ; in the gas they can go through with enormous rapidity a tremendous number of various movements of rotation and vibration, and along their free path ; and spectroscopically we can follow these movements by differences in the phenomena observed. We get a solid or liquid condition, and a continuous spectrum ; we get the most tenuous gaseous condition, and then the phenomenon is changed, and the spectrum consists of a single line. The present point is this, that, so far as the visible spectrum goes, it is possible by working with a gas at low pressure, and not too high temperature, to get a spectrum from any gas or vapour of only a single line. As we increase the density, and thus increase encounters ; or increase temperature, and thus increase the energy of each encounter, so does the spectrum get more and more complicated.

9. *In the case of metals there are two different ways in which this complexity comes about.*

We may picture to ourselves the particles cooling and losing their energy, as we get further from the source of supply ; we see that the nearer the particles are to the centre the more they bang about and the more lines we

K 2

get in the spectrum. It is important to notice that vibration once begun always goes on; it never gives *place* to others, although it may give *rise* to others; so that we get the largest number of lines in the centre, where the particles are closest together.

Here we have specially to refer to the fact that the way in which the complicated spectrum is built up varies in different substances. Plate VII. reproduces a photograph of the spectrum of aluminium and calcium compared with that of the Lenarto meteorite. The spectra of calcium and aluminium differ generically from that of the meteorite. I want to draw attention to the thick or winged lines we get in the case of aluminium and calcium. These spectra are good specimens of those which give a more brilliant spectrum by thickening the lines, while the elements in the meteorite afford good examples of those which produce a brilliant spectrum by increasing the number of their lines.

10. *When low temperatures are employed, important difference in kind is generally observed between the spectra of metals and those of metalloids, taken as a whole.*

Spectroscopically it is more easy to define the difference between these two great classes of elements than the chemists would imagine. A portion of the spectrum of the carbon vapour always present in the electric arc affords as good a representation of the spectrum of a metalloid as anything I can give. It is rhythmic. It is a " fluted " spectrum.

11. *Many phenomena observed when the molecular vibration is brought about by electricity seem to indicate that in some cases all, and in others only some, of the molecules are affected.*

When the jar spark is used with gas in a Geissler tube at low pressure, it is pretty certain that all the molecules of the included gas are affected, although, if the tube be of irregular figure or contain capillary parts,

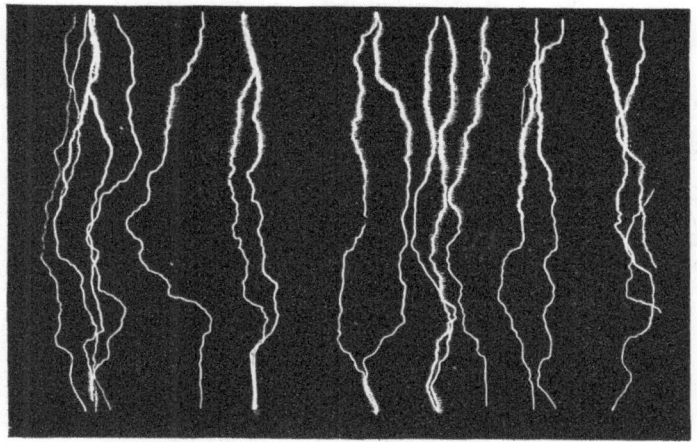

F IG. 40.—Copy of Professor Tait's photograph of the spark in air.

in all probability all the molecules will not be secondarily affected. But in almost all cases in which the spark is used without the jar, except when the pressure is very low, all the molecules do not seem to be affected. Why this should be so is seen clearly by an inspection of the beautiful photographs which Professor Tait has obtained of the spark in air.

Salet, in his beautiful researches on hydrogen, has shown that if such a spark can be got to pass through hydrogen at high pressure, the lines are thin. If we take the general view that ordinary oxygen is diatomic, we cannot understand the formation of ozone, unless we assume that only a small number of the molecules come under the influence of the spark.

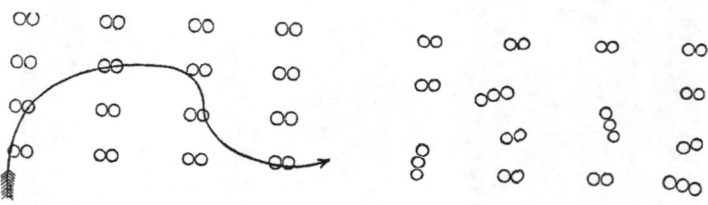

FIG. 41.   FIG. 42.

The accompanying rough illustrations will explain my meaning. On Fig. 41 we see a spark having sixteen diatomic molecules of oxygen; in Fig. 42 we see the production of these triatomic molecules of ozone as a result.

In the above cases we have assumed the molecules traversed by the spark to be all alike, to begin with, but there are phenomena observed when mixtures of gases and acid vapours are dealt with which merit much more attention than they have received up to the present time. To take an instance. In a mixture of mercury, vapour, and hydrogen, by varying the tension of the current and the pressure of the gas, it is possible to get the mercury spectrum alone, or the hydrogen spectrum alone, or both combined.

12. *Particles, the amplitudes of vibrations of which may either be so slight that no visible light proceeds from them, or so great that they give out light of their own, absorb light of the same wave-length and of greater amplitude passing through them.*

Here we are in presence of one of the grand generalizations of this century, with which the names of Stokes, Ångström, Balfour Stewart, and Kirchhoff will for ever be associated.

Consider how beautiful this statement is when we regard it in the light of its teaching with regard to molecular motions. We throw sodium into a flame and get a yellow light; we place it on the poles of our electric lamp and render it incandescent, and its light is rich yellow.

We have similarly incandescent sodium outside the sun, through which incandescent sodium the rays of sunlight pass outwards towards the earth, and we may have non-luminous sodium vapour in a test-tube; through which we may throw the white light of an electric lamp. The vibration of the sodium vapour in the case of the sun would be competent to give us the complete bright line spectrum of sodium were the interior sun abolished and only the vaporous envelope left, while the vapour of the tube is invisible in the dark. The passage of the light in both cases, however, is accompanied by absorption, and, instead of bright lines, we obtain dark ones.

Our knowledge of the elements existing in the sun and stars depends entirely, as we have already seen, upon the principle first suggested by Stokes, that particles are set swinging when waves pass through them with the particular rate of vibration which they affect.

13. *Those elements which increase the complexity of the spectrum by widening their lines, most easily produce the phenomena of absorption.*

This is another remarkable fact connected with the foregoing. A thin dark line is observed in the centre of the thick bright lines ; this is due to the absorption by the rarer cooler vapour lying outside the interior hotter vapour. This is almost invariably observed in the substances giving us the lines thickening, while iron and the allied metals does not give us any such reversal. It is well to see if one can group facts together. That is the first business of a man of science. It is extraordinary that in all the substances I have yet examined, the question of specific gravity decides not only whether the substance should have its spectrum complicated by thickening or increasing its lines, but whether such reversal shall be easily obtained. The specific gravity of iron is high. In the case of aluminium, magnesium,

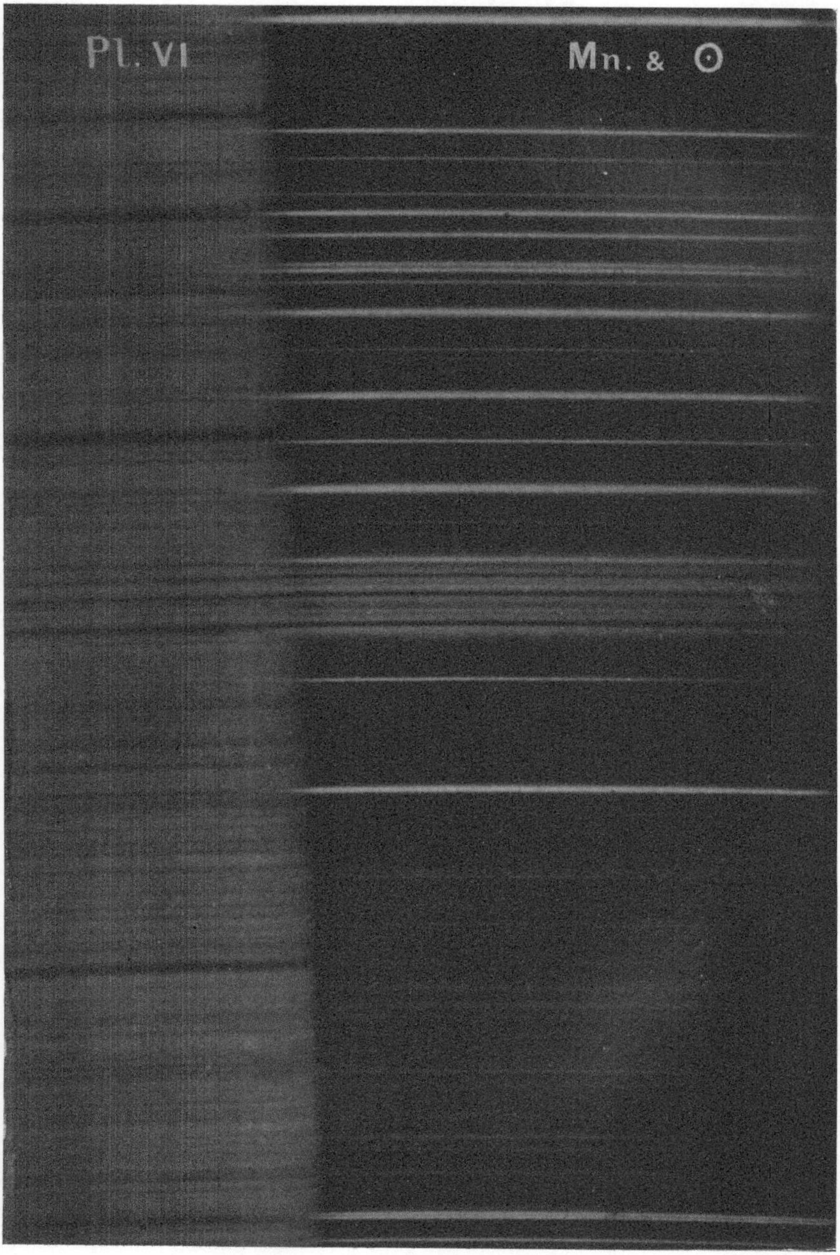

Pl. VI   Mn. & ☉

sodium, and others where this is low, we have the widening of the lines and the easy reversal.

14. *In those cases in which the light of winged lines is absorbed by the cooler exterior vapour, the shorter lines are not reversed.*

Some examples may be given. In the spectrum of coal gas, dissociated by an inducted current, only the F line of hydrogen is reversed. In the spectrum of sodium-vapour produced by the passage of the voltaic arc only D is reversed, all the other lines are bright.

In the spectrum of manganese-vapour produced in like manner, of the four adjacent lines in the violet only three are absorbed, as shown in Plate VI.

A corollary of this proposition, so far as the manganese experiments is concerned, is that the molecular motion of the vapour of manganese in the sun, is much greater than that in the cooler portions of the arc, as may be seen by reference to the solar spectrum shown in the same plate for purposes of comparison.

We find all the lines in question reversed in the latter spectrum, hence the molecular motion in the sun is greater than that in the core of the arc.

15. *A compound particle—that is a particle known to consist of two distinct elements—has a vibration which is as peculiar to itself as the vibration of a particle of an element is peculiar to itself.*

Thus to begin, with a definite example, the salts of strontium have each a distinct spectrum. Take the particle of

N_2O_4. The absorption spectrum of this gas, or rather mixture of gases, is very complex ; its particles have a vibration quite of their own. Now it is a gas which it is perfectly easy to dissociate. It is easy to turn it from N_2O_4 to NO_2. We introduce a new spectrum. These facts—and they might easily be multiplied—show then that a compound particle is a perfectly distinct physical thing, with vibrations, rotations, and free paths of its own. There is no apparent connection between the vibrations of a compound particle and those of any of the substances which make up that compound particle.

16. *In the case of metalloids, and compound gases containing them, the spectrum to a large extent depends upon the thickness of the vapour through which the light passes, and often, if not invariably, the absorption increases towards the red end as the thickness is increased.*

Here is one of the points of the most extreme theoretical importance, and one about which least is known. There is a statement in Prof. Maxwell's book, that if we take a metallic vapour and employ a great thickness of it, we shall get from it the same spectrum as from a small thickness of great density. This is Prof. Maxwell's statement. I venture to think that it requires much qualification, for in questions of thickness the spectroscope can offer the physicist a million of miles or a millimetre to work with, and one would think that such a difference should be enough.

If I take a tube with a bore of the size of the lead

in a pencil, enclose some hydrogen and render it incandescent, we see a line of a certain thickness, with a certain pressure. Looking through the sun's coronal atmosphere in an eclipse, we pierce seven or eight hundred thousand miles of hydrogen gas. The thickness of the line is the same. Various thicknesses of sodium vapour do not alter the thickness of the lines, provided that thickness is the only variable condition. But if we pass from metals to the metalloids, then the statement seems more justified. There is considerable interest attached to the question whether there is or is not any chlorine in the sun's outer atmosphere. I have endeavoured to settle this question by contrasting the absorption chlorine spectrum with the solar spectrum; different thicknesses of chlorine have been employed, and the spectrum becomes much more decided with each change of thickness. It seems that, if we take the metalloids, the absorption of a small thickness often takes place in the violet portion of the spectrum.

Now can these results be harmonised ? Here I acknowledge we tread on very difficult ground, and with our present knowledge it would be perhaps best to say nothing ; but I am not sure that this would not be scientific cowardice, so I will ask, under all reserve, whether the following explanation may not be a probable one ? With metallic vapours the lines, though not widened by thickness as they are widened by great density, are certainly darkened, but all the lines are not visible—only the longest, generally. Now if we assume that the fluted spectrum of the metalloids is really, even where it appears continuous, built up of lines,* then the darkening of these lines by greater thickness will not only make those darker that we see with a small thickness but bring others into visibility; and if this goes on till we have a very great thickness we may have an immense difference in the appearance of the spectrum. In short it would seem that there is a much closer connection between the stronger and feebler lines in the fluted spectra of the metalloids than there is between long and short lines in the case of the metals.

17. *In encounters of dissimilar molecules the vibrations of each are damped.*

We saw in 8, that the complexity of the spectrum of any substance increased with the number of encounters among similar molecules. The sympathetic encounter seems to fill the molecule, as it were, with fresh energy of the spectrum producing kind, and hence it is that the brilliancy increases either with increased density or temperature. But if we operate upon a mixture of dissimilar molecules, then experiment shows that unsympathetic encounters deprive both molecules of a part of this energy. If we confine our attention to any one of the constituents, then each increase in the quantity of another is followed by a dimming of the spectrum of the first.

* Thalén's beautiful researches on the spectrum of iodine quite bear out this view.

18. *If we are to hold that the lines, both "fundamental"* *and "short," which we get in a metallic spectrum, are due* *to encounters, then, as neither the quantity of the encounters* *nor their quality is necessarily altered by increasing the* *thickness of the stratum, the assumption that a great* thickness *of a gas or vapour causes its radiation, and* *therefore its absorption, to assume more and more the* *character of a continuous spectrum as the thickness is* *increased, seems devoid of true theoretical foundation.*

To test this point I made the following experiments :—

1. An iron tube about 5 feet long was filled with dry hydrogen ; pieces of sodium were carefully placed at intervals along the whole length of the tube, except close to the ends. The ends were closed with glass plates. The tube was placed in two gas-furnaces in line and heated. An electric lamp was placed at one end of the tube and a spectroscope at the other.

When the tube was red-hot and filled with sodium-vapour throughout, as nearly as possible, its whole length, a stream of hydrogen slowly passing through the tube, the line D was seen to be absorbed ; it was no thicker than when seen under similar conditions in a test-tube, and far thinner than the line absorbed by sodium-vapour in a test-tube, if the density be only slightly increased.

Only the longest "fundamental" line was absorbed.

The line was thicker than the D line in the solar spectrum, in which spectrum all the short lines are reversed.

2. As it was difficult largely to increase either the

temperature or the density of the sodium-vapour, I made another series of experiments with iodine-vapour.

I have already pointed out the differences indicated by the spectroscope between the quality of the vibrations of the " atom " of a metal and of the " subatom " of a metalloid (by which term I define that mass of matter which gives us a spectrum of fluted spaces, and builds up the continuous spectrum in its own way). Thus, in iodine, the short lines, brought about by increase of density in an atomic spectrum, are represented by the addition of a system of well-defined "beats" and broad bands of continuous absorption to the simplest spectrum, which is one exquisitely rhythmical, the intervals increasing from the blue to the red, and in which the beats are scarcely noticeable.

On increasing the density of a very small thickness by a gentle heating, the beats and bands are introduced, and, as the density was still further increased, the absorption became continuous throughout the whole of the visible spectrum.

The absorption of a thickness of 5 feet 6 inches of iodine-vapour at a temperature of 59° F. gave me no indication of bands, while the beats were so faint that they were scarcely visible.

19. *On the whole, certain kinds of particles affect certain parts of the spectrum.*

Take the bright lines of the metals ; if we were to mix together all the known metals in the sun, make a compound which should consist of all of them, put it into the lower pole of an electric lamp and photograph the spectrum, then we should find the majority of the lines would be in the violet end of the spectrum, scarcely any in the red end. That is the reason why the spectrum of the sun, which contains so many of the metals, is so complicated in the violet. If we combine a metal and a metalloid, we find, in many cases at all events, that the vibrations will lie in the red end of the spectrum ; we shall also find that there is a connection between the atomic weight of the metalloids and the region of the spectrum in which their lines appear under similar conditions.

We have, in fact, simple particles and short waves, compound particles and long waves. Nor is this all. In many cases we find both ends of the spectrum, and in many cases the more refrangible end only, blocked by continuous absorption. This occurs so often in absorption spectra that one is led to suspect that it is due to some arrangement of particles.

20. *Some of the vibrations are very closely connected with others, as evidenced by repetitions of similar groups of lines in different parts of the spectrum.*

Here we are brought face to face with a revelation of the vibrations of particles, which, if I am not mistaken, will be made much of by the mathematical physicist in the future.

I will content myself by giving two or three striking instances, first noticed by Mascart. We shall see that the longest line is at work in all of them.

In sodium we may say that the longest line is double; I refer to D′ and D″.   All the lines are double.

In magnesium the longest line is a triple combination. This is repeated exactly in the violet.

In manganese we may almost say that the same thing happens, but the phenomenon is much more absolute in the case of those particles such as sodium and magnesium, which, on other grounds, I suspect to be of the simplest structure.

21. *Our knowledge of the vibrations of particles will be incomplete until the vibration is known from the extreme violet (invisible) to the extreme red (invisible).*

In the meantime great help may be got from inferences, and, in the case of metalloids at low temperatures, from the position of their continuous absorption : and it is a question whether light may not be thus thrown upon the opacity of some solid substances and the transparency of others.

I think it not too much to say that already, in the case of some gases and vapours which are apparently transparent, it is as certain in some cases that their absorption is in the ultra red, as it is certain that in the case of others the absorption is in the ultra violet.   And further, it can scarcely be that this absorption is not of the continuous or fluted kind—in other words, that no gas is "atomic" in the chemist's sense, except when subjected to the action of electricity, or, in the case of hydrogen, to a high temperature.

CHAPTER V.

LONG AND SHORT LINES.

§ 1. *First Glimpses.*

FROM the time of Wheatstone's first experiments, when in 1835 he stated that in the case of the spectrum of an electric spark taken between two metallic poles, if the poles consisted of two different metals the spectrum contained the lines of both metals; down to the researches of Stokes, Miller, and Robinson in 1862 ; there is no reference, so far as I can find, to any localization of light in any portion of the *breadth* of the spectrum.

In the case of the spark taken between two poles, in air, the spectrum is generally one in which the lines of the two vapours and of air are blended together, all the lines running across the field.

But under certain conditions this is not so. Thus Stokes,* who used the spark itself instead of a slit, remarked that the metallic lines are "distinguished from air lines by being formed only at an almost insensible distance from the tips of the electrodes, whereas air lines would extend right across."

* Philosophical Transactions, vol. clii. 1862, p. 603.

L

Miller,* who used a slit and a spark close to it, referring to his photographs of electric spectra, remarks,
"the marginal extremities of the metallic lines leave a
stronger image than their central portions," and the
extremities of these interrupted lines he terms "dots."

The accompanying woodcut of the spectrum of cadmium as photographed by him will show the expres-
siveness of the word.

On the same subject Robinson † writes, "At that
boundary of the spectrum which corresponds to the
negative electrode (and in a much less degree at the
positive) extremely intense lines are seen, . . . which,
however, are short."

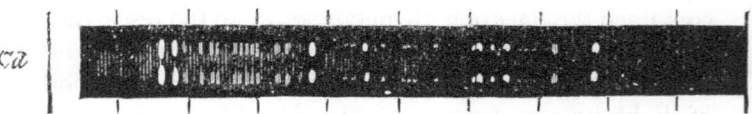

Cd

FIG. 43.—Copy of Dr. Miller's Spectrum of Cadmium, showing the "dots."

Thalén (though he also did not adopt the method
used by Dr. Frankland and myself in and since 1869
referred to at length in Chapter II.,) observed this
localization to a certain extent, doubtless on account of
the long collimator which he employed.

He remarks ‡ :—Il y a aussi des raies brillantes qu'on
n'observe que dans des cas exceptionnels, comme, par

* *Op. cit.* p. 877.
† *Op. cit.* p. 947.
‡ "Mémoire sur la détermination des longueurs d'onde des raies métalliques," p. 12, printed in the Nova Acta Regiæ Societatis Scientiarum Upsali-
ensis, ser. iii. vol. vi. Upsala, 1868.

exemple, quand la quantité de la substance soumise à l'expérience est très-abondante ou quand l'incandescence devient très-vive. Ces raies qui se présentent ordinairement aux bords du spectre sous la forme de points d'aiguille, même quand les autres raies du métal forment des lignes continues en travers du spectre, ont été representées sur la planche par des lignes très-courtes."

§ 2. *Solar Evidence.*

I may as well at once confess that when I began spectroscopic work this literature was unknown to me, and that it was to the sun that I was indebted for my first information on the subject. In 1869 I was employing all my spare time on observations of the chromosphere, and soon found that the vapours of magnesium, iron, and other metals, were sometimes injected into it.*

On the 14th of March, 1869, I observed that when the magnesium vapour was thus injected the lines of that metal did not at all attain the same height.

Thus of the *b* lines, b^1 and b^2 were of nearly equal height but b^4 was much shorter.

Next I found that of the 450 iron lines observed by Ångström, only a very few were indicated in the spectrum of the chromosphere when iron vapour is injected into it.

Dr. Frankland and myself were enabled at once to connect these phenomena, always assuming, as required by our hypothesis,† that the great bulk of the absorp-

---

* Proc. Roy. Soc. vol. xvii. p. 351.　　　† *Ibid.* p. 90.

tion to which the Fraunhofer lines are due takes place
below the chromosphere by reasoning that, as in the
case of hydrogen and nitrogen, the spectrum became
simpler where the density and temperature were less.

To test the truth of this assumption by some laboratory experiments, we took the spark in air between
two magnesium poles, so separated that the magnesium
spectrum did not extend from pole to pole, but was
visible only for a little distance, indicated by the atmosphere of magnesium vapour round each pole.

We then carefully examined the disappearance of the
b lines, *and found that they had behaved exactly as they
do on the sun.* Of the three lines the most refrangible
was the shortest.* In fact, had the experiment been
made in hydrogen instead of in air, the phenomena
indicated by the telescope would have been almost
perfectly reproduced ; for each increase in the temperature of the spark caused the magnesium vapour to extend
further from the pole.

The two annexed woodcuts (figs. 44 & 45), copied from
photographs, will show in detail the appearance presented
when the jar-spark passes (1) between the poles of *zinc*
and *cadmium*, and (2) between *cadmium* and *lead*, and
the image is thrown on the slit. It will be seen that in
the case of these metallic vapours (and it is true of all
others that I have yet observed) the lines, as in the
before-mentioned case of the triple line (*b*) of magne-

---

* The same phenomena were observed when the spark was taken in air
between magnesium poles in a Geissler tube in which the pressure was
gradually diminished.

sium, are of unequal length, and that in the new method of observation the lines in the spectra of the two metallic

Violet   Red

FIG. 44.—Copy of a photograph of the spectrum of the spark between poles of zinc and cadmium, showing the separation of the three spectra.

vapours and of the air are separated in the clearest and most convenient manner, the air lines going right across,

Violet.   Red.

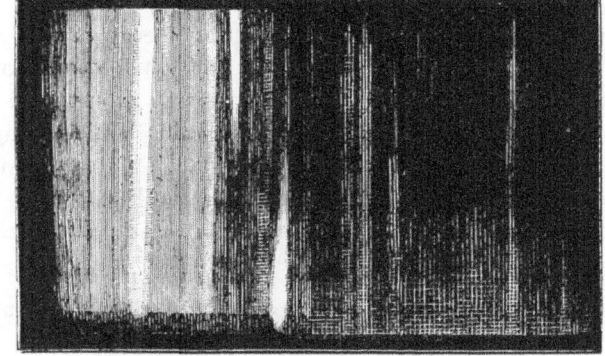

FIG. 45.—Spectrum of Cadmium and Lead.

and the lines of the metallic vapours extending to

greater or less distances from each pole and in some cases overlapping.

Many of these spectra have also been observed when the metals were enclosed in tubes and subjected to a continually decreasing pressure. *In all these experiments it was found that the longest lines invariably remained visible longest.*

In the case of zinc the effect of these circumstances was very marked, and they may be given as a sample of the phenomena generally observed. When the pressure-gauge connected with a Sprengel pump stood at from 35 to 40 millimetres, the spectrum at the part observed was normal, except that the two lines 4924 and 4911* (both of which, when the spectrum is observed under the normal pressure, are lines with thick wings) were considerably reduced in width. On the pump being started these lines rapidly decreased in length, as did the line at 4679,—4810 and 4721 being almost un-affected; at last the two at 4924 and 4911 vanished, as did 4679, and appeared only at intervals as spots on the poles, the two 4810 and 4721 remaining little changed in length, though much in brilliancy. This experiment was repeated four times, and on each occasion the gauge was found to be almost at the same point, viz. :—

> 1st observation, when the lines 4924 and
> 4911 were gone the gauge stood at   30 millimetres.
> 2nd   „   „   „   29   „
> 3rd   „   „   „   29   „
> 4th   „   „   „   31   „

> * Thalén's scale as given by Watts.

A rise to 34 millimetres was sufficient to restore the lost lines.

It soon became clear that the light given out by the discharge depended upon the amount of vapour lying between the poles, and that if both poles were composed of equally volatile metals, or the same metal, the bridge was formed by an equal, or nearly equal, amount of vapour lying round each pole ; hence, supposing that the vapours do not intermingle, it follows that the longest line can only be half the length of the actual distance between the poles.

When, however, the poles are unequally volatile, the bridge appears to be formed entirely of vapour from the most volatile pole ; hence the longest line can extend almost or quite across the space from pole to pole.

On account of these observations aluminium was used for the poles in the experiments described, as it was found that that metal was extremely refractory in the spark, *i.e.* that all its lines in the most visible portions of the spectrum were very short—the vapour which extended above the short line region being practically capable of giving only the two lines of aluminium which fall between H_1 and H_2.

There are many phenomena in connection with these observations which are well worth study ; for instance, in a case where the spectrum of copper was examined with a plumbago point opposite to the copper pole, the effect of the former was shown by the remarkable way in which the copper lines were shortened. Even when the poles were almost touching, the copper lines were confined to the copper pole, and did not extend across the spectrum.

§ 3. *Researches Opened Out.*

Since it appeared that the purest and densest vapour alone gave the greatest number of lines, or, in other words, that the truly complete spectrum of an element is alone to be obtained upon the metallic pole itself where the vapour is densest and purest, it became of interest to examine the spectrum of a compound consisting of a metal combined with a non-metallic element.

A number of experiments was therefore made, in which the metallic spectra were compared with those given by the same metals when combined with chlorine.* It was found in all cases that the difference between the spectrum of the chloride and the spectrum of the metal was :—*That under the same spark-conditions the short lines of the metals were obliterated in the spectrum of the chloride, while the air lines remained unchanged in thickness.*

Changing the spark-conditions by throwing the jar out of the circuit, this change was shown in its strongest form, the final results being that only the very longest lines in the spectrum of the metal remained.

In the case of elements with low atomic weights,

* The compounds thus experimented on were as follows, the jar being used :—Li Cl, Na Cl, Mg Cl_2, Zn Cl_2, Sr Cl_2, Cd Cl_2, Ba Cl_2, Pb Cl_2, and $Al_2 Cl_6$. For details of these experiments see *Phil. Trans.*

combined with one equivalent of chlorine, the number
of lines which remain in the chloride is large—over 60
per cent., for instance, in the case of Lithium and 40
per cent. in that of Sodium. While, on the other hand,
in the case of elements with greater atomic weights,
combined with two equivalents of chlorine, we get a
much smaller number of lines remaining—8 per cent.,
for instance, in the case of Barium, and 3 per cent. in
the case of Lead.

These interesting results suggested the extension of
the research into the examination of the long and short
lines of the metal visible in various salts. This work
will be fully dealt with in subsequent chapters.

As in the experiments already recorded the presence
of the chemically combined metalloid had its effect
upon the spectrum of the metal, it seemed desirable to
see what mechanical mixture would do.

Experiments were therefore made with mechanical
mixtures. How the results obtained in this inquiry
may in the future help us to make spectrum analysis
quantitative is stated in Chapter VIII.

It must not be forgotten that in the atmosphere of
the sun we have the most stupendous case of mechanical mixture which can be conceived. It seemed desirable, therefore, that the lines reversed in the solar
spectrum should be examined from this point of view.

I shall attempt to show in Chapter X. how our knowledge of the sun has been increased by this examination.

CHAPTER VI.

ON THE SPECTRA OF SALTS.

IT was stated in Chapter V. that we get as a result of the experiments there detailed the following facts established :

(1) When a metallic vapour is subjected to admixture with another gas or vapour, or to reduced pressure, its spectrum becomes simplified by the abstraction of the shortest lines and by the thinning of many lines.

(2) When we use metals chemically combined with a metalloid—in other words, when we pass from a metal to one of its *salts* (I used chlorine)—only the longest lines of the metal remain in the spectrum of the chloride—the number being large in the case of elements of low atomic weight and small in the case of elements of high atomic weight, and of twice the atom-fixing powder of hydrogen.

In the present chapter I propose to show what has been done in the way of following up these preliminary results, as a preparation for obtaining those referred to in the next chapter. This, therefore, is a parenthetical one.

In the first place several series of salts were used in which the atomic weights varied :

1st. In each series.

2nd. In the associated elements in each series.

With this view the spectra of the Fluorides, Chlorides, Bromides, and Iodides of Lead, Strontium, Barium, Magnesium, Sodium, and Lithium were observed.

The method given in Chapter II. was used. The spark was taken in air without a Leyden jar.

§ 1. *Lead.*

In the case of the lead salts it was found that the Fluoride gave the most and the Iodide the least complicated spectrum of the metal.*

---

* The details of the experiments were as follows :—

Plumbic Fluoride, Pb F_2.—The eleven longest lines of the following wave lengths, from Thalén's observations, 4167·5, 4246·0, 4386·5, 5163·0, 5372·0, 4523·5, 5546·0, 5607·0, 6040·0, 6059·0, and 6452·0, were seen.

Of these it is to be remarked that 4246·0 and 4386·5 are only seen for a short time when the spark first begins to pass, 6452·0 is very faint, and 5523·5 very short; so that practically the spectrum contains but seven distinctly visible lines.

Plumbic Chloride, Pb Cl_2.—On observing the spectrum of this salt, it is found to have been simplified in the following manner :

The lines left are 4167·5, 5163·0, 5372·0, 5523·5, 5546·0, 5607·0, 6040·0, 6059·0, and 6452·0, nine in number; 5523·5 has become excessively short, and 6452·0 rather brighter than it was in the fluoride.

Plumbic Bromide, Pb Br_2.—4167·5 still maintains its brilliancy undimmed, 5163·0, 5372·0, 5546·0, and 5607·0 remain; 5523·5 is just distinguishable as a dot on the pole, but 6040·0, 6059·0, and 6452·0 are completely lost, the spectrum thus being reduced to five lines.

Plumbic Iodide, Pb I_2.—4167·5 is little or not at all altered in appearance, 5163·0, 5372·0, 5607·0 yet remain; 5546·0 has become a dot, and faint indications of 5523·5, in the same state but much fainter, are visible.

The composition by weight of the compounds used are as follows :—

Pb F$_2$, Pb to F . . . . 1·0 to 0·18
Pb Cl$_2$, Pb „ Cl . . . . 1·0 „ 0·34
Pb Br$_2$, Pb „ Br . . , ... 1·0 „ 0·77
Pb I$_2$, Pb „ I . . . . 1·0 „ 1·22

Expressing these results in another way, by making the non-metallic element unity, we have :—

Pb F$_2$, F to Pb . . . . 1·0 to 5·4
Pb Cl$_2$, Cl „ Pb . . . . 1·0 „ 2·9
Pb Br$_2$, Br „ Pb . . . . 1·0 „ 1·2
Pb I$_2$, I „ Pb . . . . 1·0 „ 0·8

The lines in the spectrum of lead, then, increase in length and number as we ascend the above series, *i.e.* as the percentage weight of lead increases, as is shown in the second table. The fact may then be simply stated, that in the case of the salts of lead the complexity of the spectrum of the metal increases as the atomic weight of the non-metallic element with which it is combined decreases.

§ 2. *Barium and Strontium.*

The next metal experimented on was barium, an element of much lower atomic weight than lead, and in this respect occupying a position not very much above the mean atomic weight of the elements. It was soon found that the facts observed with lead did not completely hold with regard to barium, although they include the phenomena presented by the chloride, bromide, and iodide. Even with these salts, however, the phenomena, though the same in kind, differ somewhat in degree. For instance, the same number of metallic lines was observed in all these salts, and between the appearance of the chloride and the bromide spectrum there was no appreciable difference. In the case of the iodide, however, there was a sensible change in the direction expected from the behaviour of lead— *i.e.* the spectrum became dimmer, that is, exhibited a tendency to die out.

When, however, baric fluoride was examined, a different state of things was observed. Instead of the spectrum becoming more complex it became simpler, exhibiting in fact only the four longest lines of barium with any degree of distinctness, and these showing but little brilliancy.

Strontium behaved in the same way as barium, the falling off of the lines in the fluoride being very marked.

It will be seen from the above experiments, and from the annexed map of the strontium salts observed in air taken as a type of this group, that the general statement made for lead does not apply to barium or strontium, the chlorides, bromides, and iodides of which metals are pretty equally volatile, while their fluorides are apparently not readily volatile at any temperature which I could employ.

§ 3. *Magnesium.*

In the case of magnesium (which, as regards its chloride, bromide, and iodide, followed the behaviour of the alkali metals, to which I shall immediately refer, rather than that of barium and strontium), the fluoride exhibited the same stubborn resistance to the action of the spark. It is to be remarked that these three fluorides are non-volatile, and so infusible that even after long exposure to the current there was little or no indication of coherence ; in fact, in the case of magnesic fluoride, the salt was distinctly seen to be blown out of the cup as a cloud of dust, and when one of these particles was converted into vapour in the spark, the spectrum exhibited fragments of lines *sharp at both ends.*

On blowing a cloud of magnesic fluoride in fine powder through the spark, *b* in particular was seen as a series of three pointed lines.

§ 4. *Sodium and Lithium.*

Sodium and lithium, elements of low atomic weight, were the next metals experimented upon.

Some *sodic fluoride* was inserted into one of the aluminium cups, and opposite was placed a clean blunt aluminium point ;* the small coil and a jar were employed.

---

* Special precautions being taken to keep the poles the same distance apart in all the experiments.

On passing the spark, D only was seen. The break was then readjusted and the spark made heavier, but the result was the same. Some new and moist sodic fluoride was then placed in the cup, but the result still remained as before.

Sodic Chloride was then treated in the same way, a fresh aluminium pole being reserved for it and placed opposite to its cup. D was present and was bright ; the double line in the red once flashed in, but it was not again seen though the cup was charged and recharged with the salt repeatedly.

Sodic Bromide treated in the same way gave D, the red line being seen but once ; D, however, was brighter than before.

Sodic Iodide treated as above gave D and the red double line, which remain constantly visible, the double green line near D being also occasionally seen. D was intensely brilliant, and the salt fumed away from the pole in a dense white smoke.

A fresh attack with more powerful apparatus was then made. One of App's 6-inch quantity coils and a jar with about 224 square inches of coated surface were employed with a powerful battery of five one-pint Grove's cells. The result, however, went exactly in the same way. The Sodic Iodide and Bromide gave all the metallic lines of sodium, which were very brilliant ; whilst the Chloride gave D and the double red very bright, and stretching all across the spectrum. The Fluoride gave also D and the double red line ; but the latter only extended three quarters across the spectrum, and neither D nor the red line was so bright as they were in the chloride. Further observations showed that under certain circumstances all the lines appeared even in these latter salts ; but they were so dim as to be scarcely visible, and the fact of these compounds behaving in direct contravention to the observations with lead was established.

The following experiments were made with Lithic Iodide and Chloride in coal-gas :

Lithic Iodide gave the red line of this metal (W L 6705), extending all across the spectrum, but it was very faint, the orange line (6102) was very brilliant and about two thirds across, the blue line (4603) was very short and nebulous when seen which was only on one occasion. This differs but little from the spectrum given by the metal itself, except that the lines are in the latter case much brighter, and that the red and orange lines extend right across the spectrum, and the blue three quarters across.*

Lithic Chloride gave the red line (6705·2) thin and faint, but all across the spectrum ; the orange very bright and across the spectrum ; the blue line was also visible, but it did not extend across the spectrum.

It will be seen from the above that lithium, the least electro-positive of the alkali metals, approaches in its spectroscopic behaviour the metals of the alkaline

---

* A line at 4972 was seen also in the spectrum when the metal was used; but this has never been seen by Kirchhoff, Thalén, or any other observer except Huggins, in lithium.

There can be no doubt that it is the cæsium line, and that it is due to the presence of a trace of cæsium existing as an impurity in the lithium.

earths, strontium and barium, as it approaches them in some points of its chemical behaviour. Thus the spectrum of its iodide differs from that of the chloride as the spectrum of baric iodide differs from that of baric chloride, and not as the spectrum of sodic iodide does from sodic chloride, as might have first been supposed from the usual position, among the alkalies, assigned to the metal.

§ 5. *Flame and Weak Electric Discharge Spectra.*

In these experiments, then, in addition to information on the points already referred to, we had accumulated some enabling a comparison to be made between the spectra of salts observed with a weak electric discharge, and those observed in a Bunsen flame, which were among the first to be studied.

Some flame spectra were specially observed * in order

---

* These experiments were as follows :—

BARIUM.—*Baric Iodide.*—This salt gave the spectrum proved afterwards to be due to the oxide and the line 5534·5 very distinctly ; it coloured the flame a greenish yellow, and fused to a globule. *Baric Bromide* gave the oxide spectrum and 5534·5 with difficulty ; the spectrum was not very bright, and the flame but little coloured. *Baric Chloride* gave the same spectrum as the two salts mentioned above ; but the spectrum was much brighter, and the flame was coloured a bright pale green. *Baric Fluoride* gave scarcely a trace of the oxide spectrum, and 5534·5 was very faint indeed ; but no signs of fusion were visible, no bead being formed, and the flame was only coloured slightly and in parts.

STRONTIUM.—*Strontic Iodide,* heated on platinum wire in the Bunsen flame, gives the spectrum in the red, so well represented in Bunsen's and Kirchhoff's drawing, and the great blue line 4607·5 of the metal. *Strontic Bromide* behaved much as the iodide did, but showed more of the structure in the red. 4607·5 was also always present, and very bright, a considerable change from its appearance in the iodide, in the spectrum of which it was for a time

M

to note how the long and short lines behaved, beads of the salts being heated in the Bunsen flame on loops of platinum wire.

It was seen, then, that when spectra produced by flames are compared with those produced by the low-tension spark, the spectra of the metals in the combination are in the former case invariably more simple than in the latter, and that they are *simplified to such an extent that only the very longest line is left ;* thus :—

Baric Iodide with the low-tension spark gives five-and-twenty lines. In the flame it gives but one, and that the longest, namely 5534·5.

Baric Bromide gives five-and-twenty lines with the spark ; only one in the flame, the same longest line 5534·5.

Baric Chloride five-and-twenty lines in spark and one, 5534·5, in flame.

Baric Fluoride four lines in spark, 5534·5 alone in flame.

Again, taking the case of strontium, we find that in the case of strontic iodide thirty-two lines are observed in the spark, one alone in the flame, and that is the longest, namely 4607·5, a line by far the longest in the spectrum of strontium.

Strontic Bromide gives also thirty-two lines in the spark and but one in the flame, the same longest line 4607·5.

faint and then became brighter. *Strontic Chloride* gave the bands very brightly at first, but not so brightly after a time ; 4607·5 was fainter, but very distinct. *Strontic Fluoride* refused to give any trace either of the strontium or compound spectrum ; it is, in fact, only capable of being heated to a white heat and giving a continuous spectrum.

Strontic Chloride gives thirty-two lines in the spark, but only this one, 4607·5, in the flame.

Strontic Fluoride gives fourteen lines in the spark, but in the flame does not even give the longest line. There is, in fact, no spectrum at all due either to the metal or the compound.

It is especially to be remarked that strontic oxide furnishes us with an intermediate condition of things between the chloride and fluoride.

In this compound the band near D is the only representative of the spectrum of the compound. The longest metallic line, 4607·5, is also invisible.

§ 6. *Similarity of the Spectra of Salts Observed in Air.*

In was noticed in the earliest observations of the spectra of salts in the spark, as it had been previously noticed in the case of the flame, that after a time the spectrum was nearly the same, whatever salt had been placed in the cup. A careful study of the phenomena when the salt was submitted to the heat of the ordinary Bunsen flame, made by my assistant Mr. Friswell, taken in connection with the other branches of the inquiry, rendered it extremely probable that the band spectrum was that of the oxide.*

* The observations were made with a large Steinheil spectroscope with four prisms, such as were used for the ordinary observations of the metallic spectra with the spark. The slit was, however, wider. The salt was placed in a tangled loop on a piece of plantinum wire held in a clip. The source of heat was a small Bunsen burner.

Strontic Iodide.—When the wire was first inserted into the flame the salt fused, and the flame showed an intense yellow coloration.

The spectrum during this stage exhibited the D line very strongly, and the

These observations then, left the question something like this : The spectra of salts produced by the sparks differ from those produced by the flame chiefly in this, that we get more of the lines of the metal in the former than in the latter case.

The similarity of the band spectra observed in both

orange band of the strontic salt spectrum lying just on the least refrangible side of D ; *but the metallic line 4607·5 was invisible,* or at the best but very faint.

If the fused bead be now quickly withdrawn from the flame and held in front of a sheet of white paper, it is seen to be evolving dense violet fumes of iodine. If the bead be again thrust into the flame and the heat continued, it gradually becomes less and less fusible, and ultimately solidifies ; the yellow tinge now vanishes from the flame, and D drops out of the spectrum, the flame becoming red ; the band spectrum and 4607·5 are now very brilliant, and the solidified mass becomes white-hot and emits a continuous spectrum. If it be placed in a test-tube and treated with an acid it effervesces ; and if submitted to the action of nitric acid and starch no blue colour is produced.

Evidently, then, the iodine has been driven off, and the mass consists in all probability mainly of strontic oxide with some strontic carbonate.

It seems then that the breaking up of the compound and the volatilization of the iodine consume so much heat that the Sr I_2 never gets sufficiently heated to enable it to be volatilized ; and hence we should not see its spectrum even were the flame sufficiently hot to render its vapour luminous, and the body sufficiently stable to bear such a heat without decomposition. The fact, however, is, that what spectrum is seen is produced by the decomposing action of the flame acting either on small quantities of the Sr I_2, which do get volatilized, or, what is more probable, are mechanically carried off by the ascending currents of iodine vapour. It is to be remarked that when the bead has become infusible the spectrum begins to die out ; the orange band then only appears very faintly, and 4607·5 has utterly gone. We then have the non-volatile oxide only left in the flame, and it of course cannot give anything but the continuous spectrum due to its own incandescence.

Strontic Bromide behaves very much in the same manner as the iodide, the difference that exists depending on the greater stability of the latter compound in the flame. On removing the bead but slight fuming is observable, and there is very little odour of bromine. After the bead has been roasted for a long time it still evolves bromine, when treated with sulphuric acid and manganic oxide.

Strontic Chloride never shows any tendency to undergo decomposition ; it cases arises from the fact that the oxide is formed, the spectrum of which is observed.

How the research has been carried further, and how all the phenomena recorded have been explained by introducing the idea of dissociation of the salt by the different energies employed, will be discussed in the next chapter.

remains clearly transparent and fusible to the end, and evolves chlorine when treated for that purpose.

Strontic Fluoride was the most stable of all ; it would not fuse, gave no coloration to the flame, and obstinately retained its fluorine to the last, as was shown by the mass giving the usual reaction of fluorine when treated on a glass plate with sulphuric acid.

166

CHAPTER VII.

ON DISSOCIATION.

WE see everywhere around us an enormous number of perfectly distinct things, some of them having vital properties, some of them lifeless, motionless; but out of this apparently infinite diversity chemistry presents us with an almost perfect simplicity. It tells us that everything which exists here is really made up of one or more of only sixty-three different things; that the whole of the animal kingdom, the vegetable kingdom, the mineral kingdom is made up of only sixty-three different substances. That is a wonderful simplification, and science always simplifies.

Now we may look upon those sixty-three elements in two distinct points of view. We may consider them in their physical relations, or we may regard them in a more purely chemical aspect. If we look upon them in relation to their physical conditions, we find that amongst them are fifty-six solids, two liquids, and five gases. If we look upon them chemically, dropping all distinctions between solids, liquids and gases, we say that some of them are metals, some metalloids; and of some, it may be truly said that it is very difficult to place them exactly—to determine whether they are on the side of the metals or on the side of the metalloids. In the same way the biologist finds it absolutely impossible to put his finger upon any particular part of the organic world and say, Here the vegetable, or here the animal, kingdom begins.

These chemical distinctions then, to which we have referred, are quite independent of physical conditions. For instance, amongst the most metallic of the metals is a gas. Again, among the metals we have a liquid—mercury; so that we have a complete chain of gas, liquid and solid among the metals, although popularly the term metal is often imagined to apply only to such solids as gold, silver, and iron. On the metalloid side, again, we have gases among them—the familiar oxygen and nitrogen; we have the liquid bromine, and so on, added to other unmistakable metalloids, such as phosphorus, sulphur, carbon, and iodine, generally thought of in their solid form.

§ 1. *The Work of the Chemist.*

Now how has the chemist brought about this marvellous simplicity? One by one he has decomposed each compound body to be found throughout the whole realm of nature, he has dissociated the molecules of which each such body is composed, and has thus been able to study each constituent; when the constituent has been compound he has driven it into its elements, and so on, till he has come to the common elemental ground.

Can we be more precise than this? I think we can. I think we are safe in saying, that throughout the whole range of his work the chemist uses in the main *vibrations*. He finds the world composed of molecules in millionfold complexities, combinations and sizes, and he acts upon these molecules by vibrations. For gross molecules he finds in heat most that he wants, but when the molecules are more delicate, then electricity is called in, and electricity does for these what heat did for the others.

Let me here endeavour to make my meaning clear. Let us assume a long series of vibrations, long at one end of the series and short at the other. We know that heat consists of vibrations, we know that light consists of vibrations. Let us also think of electricity as connected with vibrations, and let us further assume these vibrations to be short. We get heat from the sun, and among these vibrations are some to which our eye is tuned. We get an immense heat vibration from the oxyhydrogen flame, but we get, practically speaking, no light. Many of the electrical phenomena with which we are acquainted take place unseen, and without heat, showing they are not long-wave phenomena; others are exquisitely visible to us, because the vibrations are within our ken; but, to get associated heat, we want pressure, or molecular complication. In fact it is *as if* we have long heat-waves at one end of a long scale, and short electricity-waves at the other, each with different functions, heat giving us with solids and liquids *visible* phenomena, because of added shorter waves ; electricity giving us visible phenomena with gases and vapours, because of added longer waves ; heat passing invisibly through gases, electricity passing invisibly through solids ; heat bringing about chemical changes in solids and liquids, electricity bringing about similar changes in the case of gases.

Now, this being so, let us assume, for the purposes of the present statement, that the molecular motion we term heat, with its long waves, chiefly affects larger molecules, that is, compound bodies, and the molecular motion electricity, whatever electricity may be, chiefly affects smaller molecules, that is, the atoms of simple substances. We shall find, in accordance with this assumption, that if a chemist wishes to reduce the millions of compound molecules in that very compound molecule a piece of ice, he applies heat, and he gets, a physical simplification, but not a chemical one, when water is produced ; a still further, and exactly similar, stage is reached when this water takes the form of steam ; but it is not till an enormous temperature, with its added short vibrations, or until electricity is employed, that the compound molecule breaks up into the simple things oxygen and hydrogen, unless another vibration of a molecule of another simple thing (or element), which shall aid in shaking them apart, is superadded.

§ 2. *The Action of Heat.*

As instances of the action of heat, I may refer to one or two familiar experiments to indicate that in a great deal of chemical action the heat vibration requisite to bring about simplification, by means of which the elemental bodies have been determined to exist as such, is supplied by the chemical action itself; it is the heat of arrested motion. In other cases we have to supply the heat artificially; but also bear this in mind, that whenever we apply heat the heat is none of our making. It also is the result of a chemical combination, as a rule. For instance, if I take some potassium and throw it into water, the potassium will instantly burst into flame. Here we have a cool metal which when put into cool water, at once takes fire in consequence of the heat of combination which has been brought about by the attraction between the potassium and the water. As the result of that heat-vibration thus introduced the water has been simplified, one of its constituent simple things, oxygen, has associated with the potassium, and the other, hydrogen, has been liberated and might have been collected in a bell jar.

Another illustration is to be got from a mixture of water and sulphuric acid. Let us pour some ether into a test-tube and insert it in a glass. When water is poured into the glass, the ether in the test-tube placed in the glass will remain as if nothing had happened. But now if we pour some sulphuric acid into the water, what happens? We get an attraction between these two things : we get a heat vibration as the result of chemical combination ; and, as the result of the heat vibration produced in that manner, the water gets hot without turning into vapour, and the ether gets hot and turns into vapour, because ether is more readily driven into vapour by such heat vibrations than water is.

Another experiment may be chosen out of many others which might have been brought forward to show the changes brought about by heat vibrations. Bichromate of potassium is, on the application of heat, instantly reduced. When I say instantly reduced, probably a few seconds may be required in order to allow the heat vibration to act; a change of colour in the solution accompanies the application of heat, communicated from an external source, in this case ; the heat of the Bunsen burner employed, however, is really an effect of chemical combination, so that we are only one stage removed.

§ 3. *Electricity.*

But not only have we heat with its long waves to bring about chemical action and its result, simplification, but, as I have said, we have another agent, electricity.

A reference to the electrolysis of water will show us at once how very different, as a rule, the action of electricity is. For this we require two tubes filled with water, and connected as to their contents by being plunged in the same mass of water; a battery, and, in each tube connected with this battery, a strip of platinum. The instant that the circuit is made complete the water is decomposed, bubbles rise from the platinum foil, which bubbles in the one case are bubbles of hydrogen, and in the other case bubbles of the other constituent of the water, oxygen. Here you see, by means not of the long waves of heat, but by means of electricity, we bring about a complete dissociation, or a complete separation of the elements of the water, a result as I have before stated, which the mere long waves of heat can never effect—a high luminous temperature—that is, added short waves—being required.

These then are instances of simplification brought about by heat and electricity. I quit this part of the subject by the remark that if we define the ultimate particles of an element as atoms; agglomerations of atoms molecules—elementary molecules when the atoms are alike, compound molecules when the atoms are dissimilar—the heat-waves generally help us to get at the molecule, and electricity helps us to get at the atom; and mark, I only say *generally*. It might be universally true if all elementary atoms were alike; but on that point we must be content to say that we do not know. I shall afterwards attempt to bring forward much evidence to show that they are vastly different. We can only study them by their vibrations; for, as Sir Wm. Thomson has calculated the atoms in a drop of water are so small, that if the drop of water were magnified to the size of the earth, the atoms would then be seen not larger than cricket-balls and not smaller than shot.

It must be clearly understood that I here refer to the true atom, and not to the atom of the chemists, the weight of which they give as the "atomic weight." It may probably turn out that this is often a molecule, sometimes a very complicated one, which great heat or electricity can split up, the latter sometimes more than once. Now before the introduction of the spectroscope into science the chemist could only be sure that he was decomposing masses of matter by the chemical analysis, first of the matter acted on, and secondly of the product of his experiment.

§ 4. *The Spectroscope and Dissociation.*

With the introduction of this instrument, however, an additional help is afforded in the study both of the quantity and quality of dissociation, and this branch of the inquiry may not only be important from the chemist's point of view, but it is found to afford most precious suggestions regarding changes of molecular structure.

Kirchhoff and Bunsen told us long ago that solids and liquids gave us continuous spectra, and that gases or vapours, that is, as I hold, the dissociated constituents of solids or liquids, gave us line-spectra. This was the first hint as to the use of the spectroscope as a detector of dissociation.

Let us see what spectra we can differentiate first of all.

If we have matter in a solid state, that is matter the molecules of which are large and are near together, agitated by the waves of heat, we get a spectrum from it of a particular kind, called a "continuous spectrum," whereas if we deal with a gas or vapour not under pressure, that is with a substance the atoms or molecules of which are smaller and further apart than in the former case, agitated by electricity, or, in some cases, by heat, we find that instead of having what is called a continuous spectrum, we have one spectrum in which the light is not continuous. We have in fact only bright lines representing a few images of the slit. This then at once enables the spectroscope to tell us the difference between the rare and the dense states of matter, quite independently of what that matter may be, and whether we use radiation or absorption as the test; since a substance with a certain molecular arrangement absorbs, as we have seen, precisely the same undulations as it gives out with the same molecular arrangement. No matter what it is, the spectroscope at once tells us whether this matter is in a gaseous or vaporous state, in which case we have lines or bands; or in a state in which the molecules are more complicated probably, as well as nearer together, when we get a more or less complete continuous spectrum.

Here then we have two distinct spectra, a banded one, giving us the individualized vibration of the smallest mass of each kind of matter, and a continuous spectrum giving us the non-individualized vibration of a more complex aggregation. Let me take two instances to render my meaning quite clear. Melted iron and melted sodium give us the same spectrum, *i.e.* a continuous one; here there is no individualizing, but the vapours of sodium and iron when electricity is employed, give us, the first, four sets of double lines, the latter perhaps a thousand lines irregularly scattered all along the spectrum. Between these stages three other spectra are observed; one special to each element and two common to all.

In fine, dealing with elementary substances the following are the spectra which (Sec. 16, Chapter IV.) have been observed :—

Line.

Fluted.

Continuous spectrum in the blue.

————————————in the red.

————————————all along the spectrum.

Here then we have already a spectroscopic discrimination of certainly two, and most probably five, different physical states of matter in the case of elementary bodies.

So much for physical differences.

How then about the chemical differences? Here the information afforded by the spectroscope is of a much closer character. The spectroscope at once enables us in the main (and I say in the main, because I have already referred to the border-land between the metals and the metalloids) to differentiate quite as sharply between metals and metalloids as it does between solids and gases.

A metal always gives a line spectrum when we employ electricity to produce the vapour, and sometimes gives us one when we employ heat.

A metalloid gives a line-spectrum when we employ electricity, but never when we employ heat. Long heat-waves in their action upon metalloidal molecules only produce bands and fluted spaces.

Here is another most important distinction.

As we can distinguish the spectrum of a metal from the spectrum of a metalloid by the appearance of the spectrum, so also does the spectroscope enable us to see a difference between the spectrum of a molecule of a compound body and an elemental molecule. Let me explain what I mean:—If we are dealing with a metallic element, we get a spectrum of a particular kind so sharply defined that when any one has once seen it, he always knows that an atom of a metal is being dealt with. In the same way when we are dealing with metalloids, the spectrum, as generally observed, is so distinct from the spectrum of a metal, that when you have once seen the spectrum of a metalloid produced by the long heat-waves, you will always be able to tell it again ; there is no possibility of mistaking it for the line-spectrum of a metal.

When we employ electricity the spectra of the metalloids present exactly the same appearance as the spectra of the metallic elements, such as iron and sodium, and it is only when we employ heat waves that those other changes to which I have referred take place.

So far we have been dealing with the elemental molecules—atoms of metals and metalloids.

Let us see what happens when we take a compound molecule. Let us, for instance, take the combination between metalloids and metals, such as some of the salts of strontium—the chloride of strontium, iodide of strontium, and so on : here we have compound molecules, that is, molecules no longer built up of one substance, but of two ; and the long heat-waves, although they can set them vibrating and therefore make them radiate light, do not shake them asunder.

We find that the spectroscope is perfectly competent to separate such spectra from all others, so that when we have once seen the spectrum of, say, iodide of strontium, we shall for ever afterwards know that such spectra are given by such a compound molecule as iodide of strontium.

The same remark applies to the compound molecules in which oxygen enters as one of the substances. Such spectra closely resemble the spectra of the metalloids, but the bands are further apart and lie nearer the violet as a rule, so that it is not difficult to distinguish them.

One word more on the fundamental difference between the spectrum of a metalloid and the spectrum of a metal on the one hand, and the spectrum of a compound on the other. The metalloid has a fluted spectrum, sometimes to be found in the central part, that is to say, in the green part, or thereabouts, of the spectrum; whereas in the case of the vapour of metals such as iron, and so on, we get bright lines only, not bands; and these lines increase in number generally toward the violet; while in the case of the compound molecules, such as iodide of strontium, to which I referred, we get a something which is half fluted spaces and bands, and half lines, but in all the cases I have examined, excluding oxides, they are limited to the red end of the spectrum.

We are now then in possession of the necessary facts to enable us to appreciate the aid the spectroscope brings, as I said before, when we wish to study the effects of dissociation.

N

Imprimis, there is the rule which is universally true, that when we drive the molecules of solids, which give us continuous spectra, into vapour—in other words, when we dissociate them, and do this in the most thorough manner—we obtain the line-spectra of their constituent element or elements.

Secondly, let us take a compound molecule, that is to say, an association of two molecules or atoms of two different chemical substances. The question of vibrations instantly comes into play ; for if the function of vibration, whether we deal with large molecules and long heat-waves, or small molecules and electricity, is to render more simple what in the first instance was compound, then we ought to get spectroscopic differences.

I propose, therefore, to discuss the question of the dissociation of known compound bodies at length. The considerations brought forward in the last chapter enables us to do this most effectively.

§ 5. *The Spectra of Bodies Known to be Compounds.*

Some of the earliest observations of this nature (1860) have been described by Kirchhoff and Bunsen.* They remark, " We have compared the spectra represented on the plate,† which we have obtained from the pure chlorides, with those produced when the bromides, iodides, hydrated oxides, sulphates, and carbonates of

---

* Translated in Philosophical Magazine, 1860, vol. xx. pp. 91–93.
† The spectra shown on the plate are those known as flame-spectra.

the several metals are brought into the following flames :—

> " Into the flame of sulphur.
>> „ „ bisulphide of carbon.
>> „ „ aqueous alcohol.
> Into the non-luminous flame of coal-gas.
> Into the flame of carbonic oxide.
>> „ „ hydrogen.
> Into the oxyhydrogen flame.

" As the result of these somewhat lengthy experiments, the details of which we here omit, it appears that the alteration of the bodies with which the metals employed were combined, the variety in the nature of the chemical processes occurring in the several flames, and the wide differences of temperature which these flames exhibit, *produce no effect upon the position of the bright lines in the spectrum which are characteristic of each metal.*

" It was found that the same metallic compound, placed in one of these flames, gives a more intense spectrum the higher the temperature of the flame. In the same flame, those of the compounds of a metal give the brightest spectra which are most volatile.

" In order to prove still more conclusively that each of the above mentioned metals always produces the same bright lines in the spectrum, we have compared the spectra represented in the plate with those produced when the electric spark passes between electrodes made of these metals.

" Small pieces of sodium, potassium, lithium, strontium, and calcium were fastened to fine platinum wires and melted two by two into glass tubes, so that the pieces of metal were separated by about 1 to 2 millims., and the platinum wires were melted through the sides of the glass tubes. Each of these tubes was placed in front of the spectrum-instrument, and by means of a Ruhmkorff's induction-apparatus,* sparks were allowed to pass between the pieces of metal inside the tube ; the spectrum thus produced was then compared with that given by a gas-flame in which the chloride of the metal was brought. The flame was placed behind the glass tube. By alternately bringing the induction-apparatus into and out of action, it was easy, without measuring, to convince ourselves that in the brilliant spectrum of the electric spark, the bright lines of the flame-spectrum were present in their right position. Besides these lines, other bright ones appeared in the electric-spark spectrum ; some of these were produced by foreign metals present in the electrodes, others arose from nitrogen, which filled the tubes after the oxygen had combined with a portion of the electrodes."†

As already mentioned in a note, the plate given with this communication shows that the spectra thus referred to by the illustrious German chemists were the flame-spectra of the elements in question.

---

* No mention is made of a jar, which doubtless was not employed.

† I shall produce evidence in the sequel to show that this explanation is probably not the correct one.

§ 6. *Mitscherlich's First Work.*

This question was advanced in 1862 by Mitscherlich,[*] and by Professors Roscoe and Clifton,[†] from whose memoirs I proceed to give extracts, and in 1865 by Diacon.[‡] Mitscherlich, in his memoir, after detailing some experiments, goes on to remark :—

" It follows from these experiments that metallic compounds do not always give a spectrum, and that in the case of those that do, the spectra are not always the same ; and, further, that the spectra are different when they are due to a metal or its combinations. We have also the right to conclude that each binary compound which gives a spectrum gives one peculiar to itself, excepting always of course when the combination is destroyed by the flame. Up to the present time we are acquainted with little beyond the spectra of the metals themselves, by reason of the facility with which the flame reduces their combinations.

" Up to the present time also it has been admitted that metals always give the same spectra with whatever they are combined.[§] As in the above experiments this was not found to be the case, it became necessary to determine whether the ordinary spectra are due to the metals or their oxides, since according to my experi-

---

[*] Ann. de Chim. et de Phys. 1862, p. 175.
[†] Trans. Lit. and Phil. Society, Manchester, 1862.
[‡] Annales de Chimie et de Physique, 4 ser. vol. vi. p. 5.
[§] This is a reference to Kirchhoff's and Bunsen's Paper just quoted.

ments all compounds which contain the metal in the form of oxide give the same spectra."

As a result of his experiments on sodium, he states that in the flames which give the line of sodium, the spectrum is due to the metal, and not to the oxide. Hence he concludes that in the case of oxides the spectrum is the spectrum of the metals.*

He then states that the new lines which had then lately been discovered without corresponding elemental lines were probably due to binary compounds.

§ 7. *Clifton and Roscoe.*

The main view in Mitscherlich's Paper, that each binary compound has a spectrum of its own, is borne out by the conclusion arrived at by Clifton and Roscoe, who remark in their Paper above referred to :—

"Kirchhoff, in his interesting memoir on the Solar Spectrum and the Spectra of the Chemical Elements, noticed in the case of the calcium-spectrum that bright lines which were invisible at the temperature of the coal-gas flame became visible when the temperature of the incandescent vapour reached that of the intense electric spark. We have confirmed this observation of Kirchhoff's, and have extended it, inasmuch as we, in the first place, have noticed that a similar change occurs in the spectra of strontium and barium ; and, in the second place, that not only new lines appear at the

---

* This opinion he corrects in his next communication, to which reference will be made hereafter.

high temperature of the intense spark, but that the broad bands characteristic of the metal or metallic compound at the low temperature of the flame or weak spark totally disappear at the higher temperature. The new bright lines which supply the part of the broad bands are generally not coincident with any part of the band, sometimes being less and sometimes more refrangible. Thus the broad band in the flame-spectrum of calcium named Ca β is replaced in the spectrum of the intense calcium-spark by five fine green lines, all of which are less refrangible than any part of the band Ca β : whilst in the place of the red or orange Ca a, three more refrangible red or orange lines are seen. The total disappearance in the spark of a well-defined yellow band seen in the calcium-spectrum at the lower temperature was strikingly evident. We have assured ourselves, by repeated observations, that in like manner the broad bands produced in the flame-spectra of strontium and barium compounds, and especially Sr a, Sr β Sr γ, Ba a, Ba β, Ba γ, Ba δ, Ba ε, Ba η, disappear entirely in the spectra of the intense spark, and that new bright non-coincident lines appear. The blue Sr δ line does not alter either in intensity or in position with alterations of temperature thus effected ; but, as has already been stated, four new violet lines appear in the spectrum of strontium at the higher temperature.

"If, in the present incomplete condition of this most interesting branch of inquiry, we may be allowed to express an opinion as to the possible cause of the phenomenon of the disappearance of the broad bands and the production of the bright lines, we would suggest that, at the lower temperature of the flame or weak spark, the spectrum observed is produced by the glowing vapour of some compound, probably the oxide, of the difficultly reducible metals ; whereas at the enormously high temperature of the intense electric spark these compounds are split up, and thus the true spectrum of the metals is obtained."

§ 8. *Mitscherlich's Second Paper.*

Two years later (in 1864) Mitscherlich, in a second communication,* expanded his views, and brought an overwhelming mass of evidence in favour of them. The methods he employed were as follows :—

The substances were heated :

1. In the flame of a Bunsen burner.

2. In the flame of coal-gas buring in oxygen.

3. In the flame of hydrogen burning in chlorine.

4. In the flame of mixtures of hydrogen and bromine or iodine-vapour burning in air or oxygen.

5. *In the case of combustible gases* they were allowed to emerge out of the middle aperture of an oxyhydrogen burner, and were burnt in air or oxygen.

In the case of non-combustible gases they were mixed with a combustible gas, such as carbonic oxide or hydrogen.

6. *In the case of solid substances* they were introduced into a tube, one end of which was connected with a Rose's

* Translated in Philosophical Magazine, 1864, vol. xxviii. p. 169.

hydrogen-apparatus; the substance was then volatilized, and the gas kindled at the other end of the tube.

7. Or the spark was taken between *poles* containing the metal or compound *in any gas ;* or between,

8. *Liquid electrodes,* in which the temperature is much lower than in 7.

From the beautiful series of researches carried on by these several methods, he concludes " *that every compound of the first order which is not decomposed, and is heated to a temperature adequate for the production of light, exhibits a spectrum peculiar to this compound, and independent of other circumstances.*"

§ 9. *Light Thrown by the Long and Short Lines.*

Some experiments of my own, commuuicated to the Royal Society in 1873, taken in conjunction with my determination of the long and short lines of metallic vapours, and the consequent simplification of the spectra by the reducton of pressure, set this question at rest, I think and in the direction indicated by Mitscherlich, Clifton, Roscoe, and Diacon ; while much light was thrown upon all the prior observations, as a consequence of which they are brought much more into harmony than at first appeared.

These experiments have been given at length in Chapter VI. I will here summarize them as a step to the work now to be discussed :—

First. I observed that whether the spectra of iodides, bromides, &c., be observed in the flame of weak spark, *in air*, the spectrum is in the main the same, as maintained by Kirchhoff and Bunsen ; but that this is not the spectrum of the metal is established by the facts, *that with a low temperature only the longest lines of the metals are present*, showing that only a small quantity of the simple metal is present as a result of partial dissociation, and that by increasing the temperature, and consequently the amount of dissociation, *the other lines of the metals appear in the order of their length with each rise of temperature.*

Secondly. I convinced myself that this is the spectrum of the oxide, because *in air*, after the first application of heat, *the spectra and metallic lines are in the main the same*, while *in hydrogen* the spectra are different for each compound, *and true metallic lines are represented according to the volatility of the compound, only the very longest lines being visible in the spectrum of the least volatile compound.*

In proof of this statement I append a drawing (Fig. 46), representing the spectra of the chloride, bromide, and iodide of strontium.   In order to avoid the introduction of the oxide spectrum, and so to secure the differentiation of the three spectra if possible, they were observed in hydrogen, which gas had been carefully treated in such a manner as to secure as far as possible the exclusion of any trace of oxygen.   It will be seen at a glance that the spectra differ not only from the spectrum given by the metal, or by its salts in air at a high temperature, but considerably amongst themselves.   In the experiments care was taken to keep the

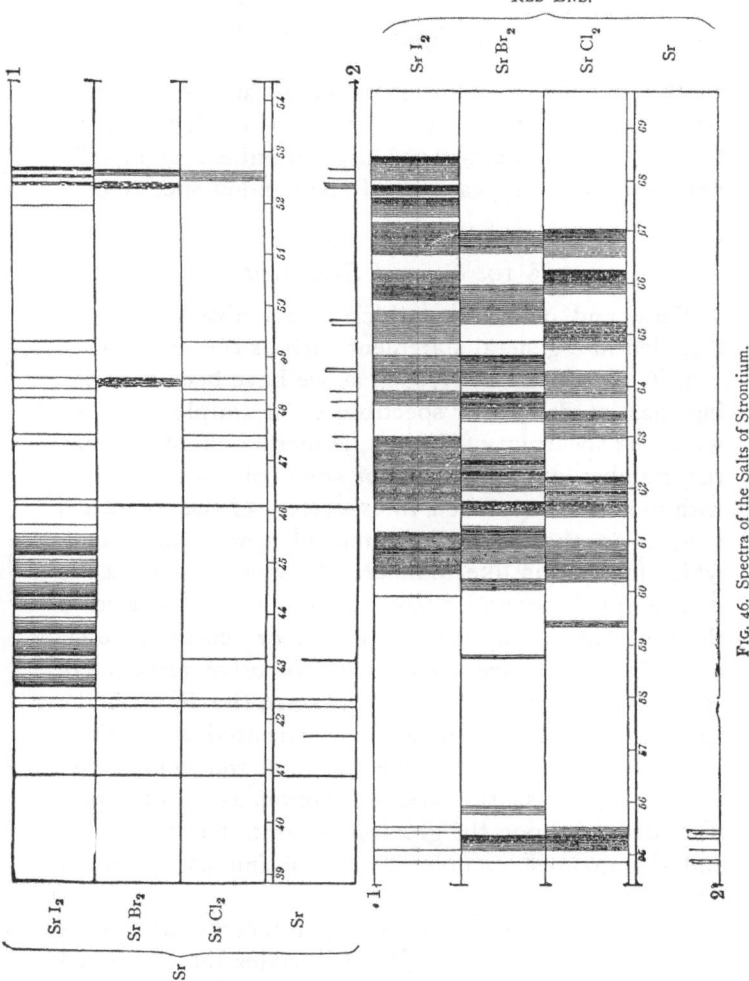

RED END.

Sr I₂   Sr Br₂   Sr Cl₂   Sr

Sr I₂   Sr Br₂   Sr Cl₂   Sr

Sr

BLUE END.

FIG. 46. Spectra of the Salts of Strontium.

spark temperature as low as possible ; and of course no jar was used, its presence in the circuit being instantly felt by the introduction of large numbers of metallic lines and the disappearance of the banded spectrum.

§ 10. *General Statement.*

These and other facts which I have observed can be included in a general statement such as the following.

1. A compound body, such as we have been considering, has as definite a spectrum as a simple one ; but while the spectrum of the simple metal consists of lines, the number and thickness of some of which increase with molecular approach, the spectrum of the compound consists in the main of channelled spaces and bands, which increase in like manner.   In short, the molecules of a simple body and a compound one are affected in the same manner by their approach or recess, in so far as their spectra are concerned ; *in other words, both spectra have their long and short lines,* the lines in the spectrum of the element being represented by bands or channelled lines in the spectrum of the compound ; and in each case the greatest simplicity of the spectrum depends upon the greatest separation of molecules, and the greatest complexity (a continuous spectrum) upon their nearest approach.

2. The heat required to act upon a compound, so as to render its spectrum visible, dissociates the compound according to its volatility ; the number of true metallic lines which thus appear is a measure of the dissocia-

tion; and as the metal lines *increase in number*, the compound bands *thin* out.*

§ 11. *The Problem of the Dissociation of the so-called Elements.*

I have now shown historically how we have been led to the conclusion that binary compounds have spectra of their own, and how this idea has been, if not established, at all events strengthened by considerations having for a basis the observations of what I have termed long and short lines.

I now proceed to show how absolutely similar observations and similar reasoning may, in the future, succeed in establishing the compound nature of, to say the least, some of the chemical elements themselves.

---

* The above statement is confirmed by the following experiments. A bead of strontic chloride was interposed between two aluminium electrodes; the induced current, without a jar, was then passed. The red-band spectrum of the oxide was very intense, and the only metallic *line* of any strength was 4607·5. The wire and bead soon became red-hot, and the latter evaporated, the spectrum disappearing. A jar of 186 square centims. coated surface was then introduced into the secondary current. The metallic lines appeared all along the spectrum, the "structure" (oxide spectrum) became fainter, and its intervals wider; the bead soon became red-hot. A jar of 467 centims. gave lines only and no structure, and one of 2214 centims. the same result, the bead remaining cold.

On using the bead as the electrode, the results were nearly the same; but the heating-effect continued when somewhat larger jars were used than the one with 362 centims., which did not show this effect with the former arrangement. When the slit was very narrow it was observed that several of the bands of the oxide spectrum broke up into masses of fine lines, exactly like those of the iodine vapour absorption-spectrum; and this remarkable resemblance was rendered still more striking by the appearance of a bead like that shown by iodine.

My attention was directed to this problem some years ago, by two perfectly distinct lines of reasoning, of which one was the foregoing.

In a Paper communicated to the Royal Society in 1874, referring, among other matters, to the reversal of some lines in the solar spectrum I remarked :—*

"It is obvious that greater attention will have to be given to the precise *character* as well as to the position of each of the Fraunhofer lines, in the thickness of which I have already observed several anomalies. I may refer more particularly at present to the two H lines 3933 and 3968 belonging to calcium, which are much thicker in all photographs of the solar spectrum, [I might have added that they were by far the thickest lines in the solar spectrum] than the largest calcium line of this region (4226·3), this latter being invariably thicker than the H lines in all photographs of the calcium spectrum, and remaining, moreover, visible in the spectrum of substances containing calcium in such small quantities as not to show any traces of the H lines.

" How far this and similar variations between photographic records and the solar spectrum are due to causes incident to the photographic record itself, or to variations in the intensities of the various molecular vibrations under solar and terrestrial conditions, are questions which up to the present time I have been unable to discuss."

In Fig. 47 I have collected several spectra copied

* Phil. Trans., vol. clxiv. pt. 2, p. 807.

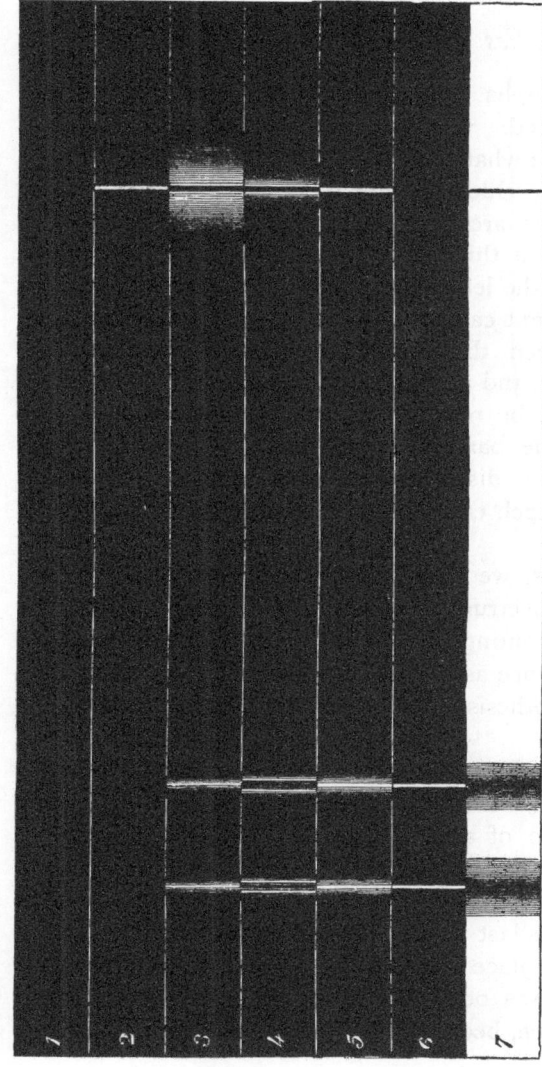

FIG. 47. The blue end of the spectrum of calcium under different conditions.

1. Calcium is combined with chlorine (CaCl₂). When the temperature is low, the compound molecule vibrates as a whole, the spectrum s at the red end, and no lines of calcium are seen.
2. The line of the metal seen when the compound molecule is dissociated to a slight extent with an induced current.
3. The spectrum of metallic calcium in the electric arc with a small number of cells.
4. The same when the number of cells is increased.
5. The spectrum when a coil and small jar are employed.
6. The spectrum when a large coil and large jar are used.
7. The absorption of the calcium vapour in the Sun.

from photographs in order that the line of argument may be grasped.

First we see what happens to the non-dissociated and the dissociated chloride. Next we have the lines with a weak voltaic arc, the single line to the right (W L 4226·3) is much thicker than the two lines (W L 3933 and 3968) to the left, and reverses itself.

We have next calcium exposed to a stronger current. It will be seen that now the three lines are almost equally thick, and all reverse themselves.

Now it will be recollected, that in the case of known compounds the band structure of the true compounds is reduced as dissociation works its way, and the spectrum of each constituent element makes its appearance.

If in fig. 3, we take the wide line as representing the banded spectrum of the compound, and the thinner ones as representing the longest elemental lines making their appearance as the result of partial dissociation, we have, by hypothesis, an element behaving like a compound.

If the hypothesis be true, we ought to be able not only to obtain, with lower temperatures, a still greater preponderance of the single line as we do ; but with higher temperatures, a still greater preponderance of the double ones as we do.

I tested this last year in the following manner :—

In the first place I may mention that I was driven to, and indeed was otherwise anxious, to employ photography—driven, because the visibility of the more refrangible lines is small—anxious, because such a permanent record of an experiment, free as it must be from all bias, is a very precious thing.

Induced currents of electricity were employed in order that all the photographic results might be comparable.

This being premised ; I may at once pass to the experiments.

To represent the lowest temperature I used a small induction coil and a Leyden jar only just large enough to secure the requisite amount of photographic effect. To represent the highest, I used the largest coil and jar at my disposal. The spark was then taken between two aluminium electrodes, the lower one cup-shaped, and charged with a salt of calcium.

In the figure I give exact copies of the results obtained. It will be seen that with the lowest temperature only the single line (2), and with the highest temperature only the two more refrangible lines (6) are recorded on the plate.

This proves that the intensity of the vibrations was quite changed in the two experiments.

When this result was communicated to the Royal Society, as will be seen from the appended note,* Professor Stokes, while accepting the new experiments as conclusive evidence of the extremely high temperature of the sun, was disinclined to go so far as to acknowledge that an element had really been dis-

---

* The note and the correspondence which arose out of it, alluded to in the text, were as follows :—

sociated.   He gave his reasons and I replied to them.
I regret to say that up to the present time the instrumental means at my disposal have not permitted me to

*Preliminary Note on the Compound Nature of the Line-Spectra of Elementary
Bodies. By J. N. Lockyer, F.R.S.* (Proc. Royal Society, No. 168, 1876.)

In a former communication to the Royal Society (Proc. vol. xxii. p. 380,
1874) I referred briefly to the possibility that the well-known line-spectra of
the elementary bodies might not result from the vibration of similar molecules;
and I was led to make the remark in consequence of the differences in the
spectra of certain elements as observed in the spectrum of the sun and in those
obtained with the ordinary instrumental appliances.

I have now clear evidence that the molecular grouping of calcium which,
with a small induction-coil and a small jar, gives a spectrum with its chief
line in the blue, is nearly broken up in the sun, and quite broken up in the
discharge from a large coil and jar, into another or others with lines in the
violet.

I say "another" or "others," because I have not yet been able to determine
whether the last-named lines proceed from the same or different molecules; and
it is possible we may have to wait for photographs of the spectrum of the
brighter stars before this point can be determined.

This result enables us to fix with very considerable accuracy the electric
dissociating conditions which are equivalent to that degree of dissociation at
present at work in the sun.

I beg permission to append the following Letter from Prof. Stokes, and my
reply :—

"March 3, 1876.

"My dear Lockyer,—You might perhaps like that I should put on paper
the substance of the remarks I made last night as to the evidence of the dissociation of calcium.

"When a solid body such as a platinum wire, traversed by a voltaic current,
is heated to incandescence, we know that as the temperature increases, not
only does the radiation of each particular refrangibility absolutely increase, but
the proportion of the radiations of the different refrangibilities is changed, the
proportion of the higher to the lower increasing with the temperature.   It
would be in accordance with analogy to suppose that as a rule the same would
take place in an incandescent surface, though in this case the spectrum would
be discontinuous instead of continuous.   Thus, if A, B, C, D, E denote conspicuous bright lines, of increasing refrangibility, in the spectrum of the
vapour, it might very well be that at a comparatively low temperature A should

put the matter to the test in the way I proposed, namely, by taking photographs of the brighter, and therefore pre-

be the brightest and the most persistent; at a higher temperature, while all were brighter than before, the relative brightness might be changed, and C might be the brightest and the most persistent, and at a still higher temperature E. If, now, the quantity of persistence were in each case reduced till all lines but one disappeared, the outstanding line might be A at the lowest temperature, C at the higher, E at the highest. If so, in case the vapour showed its presence by absorption but not emission, it follows, from the correspondence between absorption and emission, that at one temperature the dark line which would be the most sensitive indication of the presence of the substance would be A, at another C, at a third E. Hence, while I regard the facts you mention as evidence of the high temperature of the sun, I do not regard them as *conclusive* evidence of the dissociation of the molecule of calcium.

> " Yours sincerely,
> " G. G. STOKES."

> " 5 Alexandra Road, Finchley Road, N.W.,
> " March 5, 1876.

" DEAR PROFESSOR STOKES,—I was not prepared for your suggestion, as it was the abnormal and not the normal behaviour of Ca which led me to investigate it.

" D is darker than any other of the Na lines, and H in the chromosphere at the Ca level is red, while in the coronal atmosphere it is green ; *i.e.* the *least* refrangible line is developed by increase of temperature, and not the more refrangible one.*

" I am not the less grateful to you for your suggestion ; and so soon as I can obtain the use of a more powerful coil I will go over the ground as completely as I can.

" Are you quite sure that the molecular structure of the platinum wire is constant while it behaves as you say it does?

" I beg you will permit your letter and this to appear in the Proceedings. It will make my note more useful if you will.

> " Believe me very faithfully yours,
> " J. NORMAN LOCKYER."

* [The point, however, is, Which is the *most persistent* line at the respective temperatures, *i.e.* the last to disappear when the quantity of substance present is continually reduced? and Mr. Lockyer himself has shown that the line which is the most conspicuous when there is substance enough present to show several lines is by no means necessarily the most persistent.—G. G. S.]

O 2

sumably hotter, stars. But Mr. Huggins has obtained such a photograph, and although I have not seen it, he stated, at a meeting of the Physical Society, in reply to a question put by myself, that only one of the solar H lines was represented in the spectrum of *a* Lyræ. Should subsequent work confirm this result, I shall certainly regard the question as settled, and settled in the direction to which the experiments point.

Perhaps it may not be superfluous here to state the reasons which induced me to search for further evidence in the stars.

It is abundantly clear that if the so-called elements, or more properly speaking their finest atoms—those that give us line spectra—are really compounds, the compounds must have been formed at a very high temperature. It is easy to imagine that there may be no superior limit to temperature, and therefore no superior limit beyond which such combinations are possible. Because the atoms which have the power of combining together at these transcendental stages of heat do not exist as such, or rather they exist combined with other similar atoms, at all lower temperatures. Hence association will be a combination of more complex molecules as temperature is reduced, and of dissociation, therefore, with increased temperature there may be no end.

That is the first point.

The second is this :—

We are justified in supposing that our calcium, once formed, is a distinct entity, whether it be an element or not, and therefore, by working at it alone, we shall never know, even if its dissociation be granted in the future, whether the temperature produces a simpler form or more atomic condition of the same thing, or whether we actually break it up into X + Y, because neither X nor Y will ever vary.

But if calcium be a product of a condition of relatively lower temperature, then in the stars, hot enough to enable its constituents to exist uncompounded, we may expect these constituents to vary in quantity ; there may be more of X in one star and more of Y in another ; and if this be so, then the H and K lines will vary in thickness, and the extremest limit of variation will be that we shall only have H representing, say X in one star, and only have K representing, say Y in another. Intermediately between these extreme conditions of cases, we may have cases in which, though both H and K are visible, H is thicker in some and K is thicker in others.

Now, according to my interpretation of Mr. Huggins' photograph, *a* Lyræ really represents an extreme case. This, of course, requires confirmation ; but I think it a point of great interest that the extreme importance of further and careful work should have been so abundantly demonstrated by the very first photograph of a star with a comparison with the solar spectrum which has rewarded Dr. Huggins' well-known skill.

It is a matter of great satisfaction to me that the Dutch Society of Sciences has endeavoured to interest other workers in this question by offering a prize for a continuation of the inquiry. I may add, as a suggestion to those who feel inclined to take up the subject, that there are some anomalies in the spectra of lithium and magnesium which lead me to think that these metals may repay investigation from this point of view.

CHAPTER VIII.

AN ATTEMPT AT QUANTITATIVE SPECTRUM ANALYSIS.

§ 1. *Introductory.*

BOTH in France and Germany, attempts have been made to apply the spectroscope to the quantitative determination of the substances with which the spectroscope can deal ; and although up to the present time the results obtained are not such as to cause the methods to be generally used, there can be no doubt that in future, spectroscopy will render such aid to the other arts and sciences.

Of this we may be sure, that so far as practical benefit is concerned, the purer the science is to begin with, the richer it will most probably be in practical applications. In the case of the application of the abstract ideas of spectrum analysis to the arts, we have not only a promise rich so far as the arts are concerned, but, what is very much more precious to a man of science, a promise of rich results in the science itself. For, depend upon it, that as spectroscopy becomes the daily work of iron-founders, and miners, and the like, it will be found to be bristling with beautiful scientific truths in every part of the spectrum, which may be used in these practical applications of the science of optics.

In my own case I can certainly say, that if this question had not arisen as a mere matter of abstract inquiry, it is very difficult to say how it could have arisen at all. Quantitative determinations were not included in any way in Kirchhoff and Bunsen's work or ideas ; so far, indeed, as they went, it would have remained qualitative merely to the end of the chapter.

I believe that the first suggestion of a quantitative use of the spectroscope was made by the illustrious Janssen ; he was followed by Vierdort.

The method at which I have worked, differs fundamentally from theirs, and as the steps which led to it may not be without interest to some, as an instance of a practical application growing out of a theoretical inquiry, I will state them here as briefly as may be, though they are separately referred to in different parts of the book.

The generalization which those two distinguished German chemists, Kirchhoff and Bunsen, got out of their earliest spectroscopic observations—observations connected with Stokes's generalization, which dates as far back as 1852,—was an enormous advance, and a thrill ran through the scientific world when it was announced.

They showed that, although in the case of solids and liquids we always get the same spectrum, and that, therefore, *pro tanto*, spectroscopy could not help us to an analysis of anything that existed in the solid or liquid state, still, when we got the elements into a state of gas, the lines were never the same for any two substances. Here was the power, a new method of *qualitative* analysis. This work was afterwards much extended by researches in our own country. Dr. Frankland showed that in the case of hydrogen gas (and let me remark that this very beautiful abstract consideration came out of an inquiry which had for its object the illumination of our metropolis), which, of course, exists as gas at all known temperatures, though there is ample evidence that its molecular condition is not always the same, Kirchhoff's generalization was not true if the gas were at a high pressure. Dr. Frankland showed, in fact, that at a pressure of ten or twenty atmospheres the spectrum of hydrogen was as continuous as that from burning coal, or any other burning substance in a liquid or a solid state.

Here there was an apparent anomaly. Now, whenever we hear scientific men talking about an anomaly, in nine cases out of ten it is because they do not know the law. And I hope to be able to show that this anomaly was really no anomaly at all, the moment we got a higher generalization. Now what was that higher generalization? Before I state it, let me refer for one moment to some other work, done by Plücker and Hittorff, also on hydrogen, the upshot of which was that, under certain conditions—the pressure they employed being far below that employed by Frankland—the well-known lines, the one in the red and the other in the green (there are others, but we need not discuss them), were sometimes seen very thick, sometimes very thin. Plücker and Hittorff did not see exactly the cause of this variation, but that the variation existed was absolutely undoubted.

So that taking together Kirchhoff's generalization, Frankland's work, and Plücker and Hittorff's work, we had a horizon something like this. Solids and liquids give us a continuous spectrum ; the gases give us a discontinuous spectrum ; that is to say, they give us bright lines. Then Frankland's work came in.

At a very great pressure hydrogen does not give us bright lines at all ; all the bright lines are absolutely lost in a continuous spectrum ; and Plücker and Hittorff come in and teach us that, for some cause or other, which is indeterminate—it may be pressure or it may be temperature—the hydrogen lines, observed at a much lower pressure than that employed by Frankland, vary their thickness.

So much for the first state of the laboratory work. About this time the sun was being spectroscopically examined with great minuteness. In the year 1868 a new method was for the first time employed, by which we could observe the various gases which surround it ; and among these gases this very gas, hydrogen, preponderates to an enormous extent. This, of course, gave us an opportunity of studying the hydrogen of the sun.

The red line in the spectrum of the solar hydrogen was a perfectly straight line, but there was a difference between this line and the other one in the green. The latter widens as we get towards the sun ; in fact, it is trumpet-shaped.

Now, when that observation was first made, the question was how to reconcile the thickening of that line with Frankland's and Plücker's and Hittorff's work, to which I have referred. Nothing was more easy.

We know perfectly well that if we go up a mountain the pressure of the air is reduced ; and similarly we may imagine that far above the sun the pressure of the hydrogen, which practically is the equivalent of our atmospheric air, will be very much less than it is at the sun's surface ; and all we have to imagine is, that at the sun's surface the pressure of the hydrogen is very much greater than it is at a considerable distance above the sun. Take that in connection with Frankland's work, and Plücker's and Hittorff's results.

Frankland's work shows us that increase of pressure gives us a continuous spectrum ; this is an approximation to a continuous spectrum,—and the doubt which resulted from the indeterminateness, so to speak, of Plücker's and Hittorff's work, is resolved, by this solar observation, in favour of the hypothesis that in the spectrum of hydrogen, as its pressure is increased its lines widen, until at a pressure of twenty atmospheres we get a continuous spectrum.

I have already called attention to the spectrum of magnesium and of the sun's chromosphere given in Plate II. But notice, that when we observe, as we can by the new method to which I have referred, the spectrum of the magnesium vapour which surrounds the sun, as the hydrogen does, we find that the various lines of magnesium vapour do not come up to the same height.

We have three lines, two of them very long and one of them short. Now, how can we explain this? How can we square this observation, so to speak, with the observations of the hydrogen spectrum, to which I have referred?

In this way. Increase of pressure, in the case of hydrogen, widens the lines. Increase of pressure, in the case of magnesium vapour, not only widens the lines but increases their number. These are the two observations of the solar chromosphere, which form the basis of my attempt at quantitative spectrum analysis. Now it would never do to rest content with this mere interpretation of an observation of the hydrogen and the magnesium at the sun.

Laboratory experiments were therefore determined upon, and Dr. Frankland and myself saw that there were some, perfectly simple, which would enable us to see whether the hypothesis was right or wrong. All we had to do in the case of hydrogen, and what was done, was to take a tube, connect it with an air-pump, to fill the tube with hydrogen, then pump out the hydrogen, and to see whether the lines got thinner as the pressure of the hydrogen was reduced. They did get thinner.

Another series of experiments was to take a tube like the former one, but, instead of pumping hydrogen out, to pump hydrogen in until we got a great pressure, in which case the lines ought to thicken, and thicken until we got a continuous spectrum. That also we did.

But how about the magnesium vapour? That question

was also susceptible of being determined by a simple experiment. Still taking a similar tube, having the electrodes of magnesium, put these electrodes in an atmosphere the pressure of which is being constantly reduced, and then, if it be true that the magnesium spectrum gets simpler and simpler as the pressure is reduced, we shall find the magnesium lines, which we see at a pressure of one atmosphere, considerably reduced either in thickness or in number, or in both, at a pressure, say of a millimetre, or ten millimetres. That was done and the hypothesis established.

Since these first experiments a very large series of observations have been accumulated, tending to show that reduction of pressure is the thing that we must look to in these inquiries.

Now, what does reduction of pressure mean? In every room there is atmospheric air. What is atmospheric air? Let us assume that it consists of an aggregation of molecules. Suppose we reduce the pressure in a room, what would we do? We separate the molecules, and if we increase the pressure we shall increase their encounters.

The experiments I have referred to, prove that the moment we so deal with molecules, the spectrum gets more complicated, the lines get thicker and increase in number; whereas, if the molecules are separated, the lines will get thinner and the spectrum will be simpler.

And we may go very much further; we may go, in fact, as far as this :—If it were possible physically to lay hold of a single atom of any chemical substance, and by means of its spectrum observe its vibrations, we should get one line from that single atom if it were as nearly as may be in a state of rest. That we call the "fundamental line" of that particular atom, whether hydrogen, sodium, barium, chromium, calcium, or anything we choose to name. By separating the atoms of that particular chemical substance we get one line.

What do we do by bringing the atoms nearer together? We thicken the lines, or bring in more lines. We go on increasing the complexity of the spectrum, either, in the case of some chemical elements, by thickening the lines, or in the case of others—and this is an important distinction—by increasing the number of the lines. When the *molecules* are close together, we get the greatest complication of spectrum, and if we go to the other end of the series we have a single line for each chemical substance upon which we experiment.

If, therefore, when we get the atoms or molecules of a substance to separate we have a continuous scale, from a single line to the greatest complexity of spectrum, a great advance has been made. From the condition of iron, which gives us a single line, to that condition which gives us, as we know, and as we have mapped, 460 lines, the spectrum tells us of the existence of stages where we get something between one line and 460. That is quantitative analysis.

The frontispiece is a reproduction of a photograph which will show why this work was first undertaken with reference to the solar spectrum, although it has struck out from that region, and deals now with very many things apparently unconnected with the solar spectrum. The plate shows that in the solar spectrum there is an immense, so to speak, *individuality* in the lines. The lines are not all equally thick. The two thick lines in the middle are due to the absorption of calcium vapour in the sun's atmosphere, and the intermediate lines, and the lines to the right and left are, as we now know, due to the absorption, not of calcium, but of cobalt, nickel, aluminium, and other solar metals.

Some of these lines are excessively thin. I have already mentioned that, in passing from the atom undisturbed to the atom which is, so to speak, subject to reflex action, we multiply the lines in the spectrum, and thicken them. Now, when Kirchhoff and Bunsen had stated their generalization, with regard to solids and liquids giving us a continuous spectrum, and gases giving us a discontinuous spectrum, it was in reference to the solar spectrum that that inquiry was undertaken ; and in the same memoir they showed that a great many substances, with which we are perfectly familiar, are present in the sun ; hydrogen, sodium, magnesium, iron, nickel, cobalt, chromium, and the like. But they also showed that when a complete correspondence is sought between the bright lines of the vapour of any of these metals in the sun, and the dark lines in the sun's spectrum, that this very curious effect came out, that not all the lines were reversed.

For instance, in the case of iron, we have 460 bright lines. But in the sun we find about 455 dark lines corresponding, and five, let us say, do not correspond. In the case of the bright line spectrum of aluminium, only two are observed to be reversed in the sun. How was this? Kirchhoff imagined that probably it was simply due to the fact that only the brightest lines were reversed. That if certain lines were observed to be faint, then that probably those faint lines, for some unknown reason, would not be observed in the solar spectrum.

Ångström took up the question afterwards, and placed some more metals in the sun, but was unable to explain how it was that in the case of aluminium, two out of about thirty lines, let us say, were reversed, all the others absolutely leaving no trace whatever in the solar spectrum. Now it has already been pointed out that in that particular part of the solar spectrum where magnesium lines occurred, we have three dark lines due to the absorption of magnesium vapour, while when we got magnesium vapour produced not on the sun, we observed the two least refrangible lines going very much further from the source of the supply of the vapour than that other fainter more refrangible line. Now, suppose, for instance, that the pressure of the magnesium vapour in the sun's chromosphere were reduced ; what would happen ? We should not have that dark line of magnesium absorption in the sun's spectrum at all, and that is the explanation of the fact, that in the case of some substances known to exist in the sun, we do not get above ten per cent. of the lines.

I have enlarged upon this class of facts in another chapter, and as the method of obtaining long and short lines has already been referred to, I may at once proceed to point out how the quantitative result of the work comes out. We have—if there is any truth whatever in this hypothesis, and if it is true to say that there is a great deal of calcium in the sun, because the calcium lines are thick, and if it is true to say that there is not much aluminium in the sun, because all the aluminium lines are not reversed, and what are not reversed are thin—these effects instantly following. If we take a substance such as strontium, and observe its spectrum, and if we get certain lines from the dense vapour of the pure metal, and then under exactly similar conditions, instead of using the pure metal we use a chemical combination, let us say a chloride or an iodide, we ought not to get so many lines from the combination as we got from the pure metallic vapour. Now, that experiment has been tried. And it is perfectly true that under like conditions we do not get so many lines from the spectrum of any of the salts of strontium as we get from pure strontium, and that fact is not only true of strontium, but of everything else.

Now, which lines ought we to get ? The long lines or the short ones ? We ought to get the long lines, because the short lines are only given by a dense vapour. And so true is it, that we can, as it were, roughly predict the spectrum of the salts of a metal, knowing the atomic weight of the metalloid with which we chemically combine it ; provided, always, that the volatility of the compounds is equal. That is one stage. We have got

P

the same spectrum now by a chemical combination which we formerly got by a reduction of pressure.

There is no particular magic in chemical combination. Why not mechanical mixture ?

§ 2. *First Experiments.*

A special series of experiments on the spectrum of mechanically mixed metals—alloys prepared *ad hoc*— was made to test this.

A cursory examination of the spectra of some amalgams of tin and magnesium soon showed that this is the case.

For instance, I found it possible to begin with an alloy which shall only give us the longest line or lines in the spectrum of the smallest constituent, and by increasing the quantity of this constituent the other lines can be introduced in the order of their length. This reaction is so delicate that I learnt from it a thing I had not before observed, that the least refrangible line of *b*, the triple line of magnesium, is really a little longer than its more refrangible companion ; for the spectrum of magnesium was reduced to this one line in an alloy in which special precautions had been taken to introduce the minimum of magnesium.

It followed from this that not only is spectrum analysis almost infinitely more delicate than it has hitherto been supposed to be in the case of the elements in which the difference between the longest and shortest lines is least, but that in time it might become quantitative; for if the admixture of certain other bodies extinguishes the shorter lines of metallic spectra, it would seem that a series of carefully executed maps of the spectra of alloys, the proportions of the constituents of which are known, will place in our hands the means of determining (roughly it is true) by mere inspection the quantity of the sought metal present in an alloy, the composition of which *quâ* that metal is unknown.

After these preliminary experiments with amalgams, mixtures were prepared for experiment in the following manner.

§ 3. *Second Series of Experiments.*

A quantity of the larger constituent, generally from five to ten grammes, was weighed out, the weighing being accurate to the fraction of a milligramme; and the requisite quantity of the smaller constituent was calculated to give, when combined, a mixture of a definite percentage composition by weight (this being more easily obtainable than a percentage composition by volume).

The quantities generally chosen were 10, 5, 1, and 0·1 per cent.

In a few cases with metals known to have very delicate spectral reactions a mixture of 0·01 per cent. was prepared.

The larger constituent was then introduced into a small crucible (the bowl of a common clay tobacco-pipe). A tube conveying a stream of pure dry hydrogen was introduced into the mouth of the crucible, and the metal heated by a Bunsen burner.

As soon as it was melted, the metal, the spectrum of which was to be examined, was introduced in fragments, the hydrogen stream being kept up, and the heat raised if necessary until the last added metal had melted. When this had taken place, the fused mixture was agitated by rapidly shaking the crucible, or by causing the hydrogen to bubble through the melted mass.

When the mixing was judged to be complete, the mass was poured out. On cooling, a point was cut from it and placed in the spark-stand, the opposite pole being made of the metal which constituted the bulk of the mixture. Thus an alloy of 90 parts tin and 10 parts cadmium would have a tin pole opposite, and one of 90 parts lead and 10 zinc a lead pole, and so on.

It is important that each electrode of a mixture should have its corresponding electrode of the pure metal which exists in the greatest quantity in the mixture, as when the spectrum is observed the long and short lines of this constituent are seen stretching from top to bottom of the spectrum, the longest lines being continuous across, while the lines of the smaller constituent are seen *only* at the top or bottom of the spectrum, according to the place occupied by the mixture in the spark-stand.

Observations were then made of the spectrum of each specimen, and the result was recorded in maps in the following manner :—First the pure spectrum of the smallest constituent was observed, and the lines laid down generally from Thalén's map.

The mixture containing the greatest percentage of the substance whose spectrum is to be studied was then inserted in the spark-stand with its opposite electrode of the substance with which it is mixed, and the spectrum observed. This process was repeated until the lowest percentage is reached.

The series thus mapped were as follows:—

$$
\begin{array}{lll}
\text{Sn+Cd percentages of Cd} & 10, 5, 1, & 0\cdot15 \\
\text{Pb+Zn} \quad\quad\quad ,, & \text{Zn } 10, 5, 1, & 0\cdot1 \\
\text{Pb+Mg} \quad\quad\quad ,, & \text{Mg } 10, 1, 0\cdot1, 0\cdot01 &
\end{array}
$$

I may remark that these substances were used in consequence of their low fluid temperatures and of the consequent ease with which the mixtures could be made with the arrangements at my disposal.

But the maps showed that the lines disappeared as the quantity of the smallest constituent got less. Although we had here the germs of a quantitative spectrum analysis, the germs only were present, because from the existence of several "critical points," * and great variations due to other causes, the results obtained were not sufficiently close or constant for practical application.

Further researches, however, soon showed me that this method of eliminating lines from the spectrum was not the only one which might be taken advantage of.

* " Philosophical Transactions," 1873, p. 261.

§ 4. *Additional Phenomena Observed.*

After the second Paper was sent in to the Royal Society, I therefore commenced a series of observations, the object of which was to study not only the disappearance of the lines, but other general changes which might supervene; and for this purpose I mounted a micrometer eyepiece on the observing-telescope of the spectroscope. This enabled me to notice the following additional phenomena, in which a change in the lines which remained was brought about by a change of composition.

I. The lines varied in their lengths as the percentage of the element to which they were due varied.

II. Some of the lines appreciably varied in their thickness and brightness, or both, in the same way.

III. In cases where the brightness of a line was estimated through a considerable range of percentage composition by comparison with an air-line, the latter was observed to grow faint and then to disappear when the brightness of the line compared with it increased.

IV. In cases where the brightness or thickness of the line of one element was estimated by comparison with the adjacent line of the other constituent of the alloy, the point of equal brightness was observed to ascend or descend. (I used this method in order to avoid the uncertainty of micrometric measurements of the tips of the lines in consequence of their variation in length due to the unequal action of the spark.)

V. In some cases where the percentage of a consti-
stuent was so small that none of its lines were visible,
there yet seemed to be an effect produced upon the
lines of the other constituent as compared with those
of the spectrum of the same vapour of the opposite
pole.

VI. Changes in the relative lengths of a pair of lines
belonging severally to the constituents of the alloy.

These conclusions were derived from observations of
the alloys, which I had made in the manner indicated in
my second Paper ; and I saw that it would be important
to observe series in which the change of percentage
composition between the specimens was not so great,
and of which actual assays had been made.

I therefore begged Mr. C. Freemantle, the Deputy
Master of the Mint, to allow me the use of specimens
of the gold-copper and silver-copper alloys prepared
for the coinage, as in them I had exactly what the
research required—namely, ranges with small variations
and of undoubted accuracy. That gentleman, at once,
with the greatest promptitude and courtesy, acceded to
my request.

§ 5. *Experiments with Large Variations.*

Before, however, I proceed to consider the cases of
the Mint alloys, it will be well to briefly notice some
experiments with alloys in which the variations in the
proportions of constituents were greater.

As an instance of a moderately small difference, an

attempt was made with a portion of a half-sovereign to which $\frac{1}{1000}$ of its weight of copper was added. This was compared with another portion of the same coin ; but no difference was detected, possibly on account of a failure in making an alloy on so small a scale without some loss of the smaller and more oxidizable constituent.

An alloy of 5·5996 grms. lead and 0·6221 grm. silver was then made, the silver being dissolved in the lead, which was fused in a bent hard-glass tube in a current of hydrogen.

The percentage composition of this alloy was 89 per cent. lead and 11 per cent. silver. The silver lines at w. l. 5470, 5464, 5209, were recognized in its spectrum, and another line at 5401 was uncertain.

The opposite pole was composed of some supposed pure lead, and it was observed that its lines were longer than those of the lead in the alloy. A very faint line was observed in its spectrum opposite to the silver line 5290 ; but this was most probably the faint lead line 5206, as the longer silver lines did not exhibit themselves in its spectrum.

The opposite pole was now replaced by one of common sheet lead (it had at first been pure assayers' sheet), and the line in question was not stronger than in the pure lead ; hence it was almost certainly the line 5206 of lead.

In order to see how much silver was required to render its spectrum visible, and as the common sheet lead was suspected to contain traces of silver, an alloy was made which contained 0·01 per cent. of silver. In this no silver lines were visible ; and the same result appeared when 05 per cent. silver was alloyed with the lead. The same was the case when 0·11 per cent. was added ; and even the use of a large jar failed to cause the silver lines to appear.

Finally, an alloy containing 1·0 per cent. was made, and in this the silver lines appeared very distinctly, and three in number ; 5464 was the longest, 5209 next, and 5470 the shortest.

Another alloy with 0·5 per cent, was now made, and in this the two lines of silver at w. l. 5464, 5209 were at last discovered, but they were very short and faint. Continued work showed that the line 5464 dies out between ·05 per cent. and 0·2 per cent. of silver (for it was at last discovered in the ·05 per cent. alloy), and also convinced me of the extreme irregularity of the various portions of the alloy, though they had been made with the greatest accuracy which the means at my command permitted.

Experiments were also made with alloys of tin and cadmium, the latter forming 0·154 per cent. of the alloy. The longest cadmium line at w. l. 5085 alone remained permanently visible, exceedingly faint, but unchanged in length ; it had a short bright stump. When the line at w. l. 4799 was observed it appeared as a stump only, and neither 4677 nor any other cadmium lines could be found. Another alloy of tin with 1·0 per cent. cadmium showed 5085 distinct and bright, as was 4799 ; 4677 was likewise distinctly visible, and the tip of the least refrangible winged line 5377 commenced to appear.

The other winged cadmium line, 5338, could not be identified, as it is nearly coincident with a tin line. After this, with the increased percentages of 5 per cent. and 10 per cent of cadmium, the effect was mainly to lengthen and brighten the longest lines.

In order to determine whether these results would be affected by the presence of other metals in the alloy, one was made of 89·11 per cent. lead, 9·90 per cent. zinc, ·09 per cent. cadmium, and 0·85 per cent. tin ; and the cadmium line showed no appreciable difference from its appearance in an alloy containing 0·1 per cent. cadmium with 99·9 per cent. of a single other metal.

§ 6. *Gold Copper Alloys.*

In the Mint specimens with their small variations I found the same phenomena *en petit* which I was already familiar with *en grand*, and the smaller variations in the phenomena enabled me to understand them better.

I found that in the gold-copper standards an increase of a 1000th part in the gold brought the lines down, while a similar increase in the copper carried them up, *i.e.* increased the height of the vapour from the pole.

I found, on the other hand, that in the silver-copper standards an increase of a 1000th part in the silver carried the lines up, while a similar increase in the copper brought them down in the field of view, *i.e.* reduced the height of the vapour from the pole.

After registering these facts, I saw at once that all the phenomena might be explained by assuming a change of volatility ; by assuming, in fact, that alloys differing a 1000th part are different physical things, and that the spark acts upon the alloy as a whole as well as upon each vapour separately.

Thus in the cases referred to, in which copper is common to both, we find the melting-points to be as follows :—

> Gold .   .   .   1200° (Pouillet).
> Copper between 1200° and 1000°, precise point not
>   determined.
> Silver .   .   .   1000° (Pouillet).

And the intermediate position which copper occupies at once explains the different actions on its lines brought about by the addition of gold and of silver.

Mr. W. C. Roberts, the chemist of the Mint, was good enough to join me in a detailed examination of the gold-copper alloy.   In our experiments, unknown alloys and check pieces of known composition having been arranged on a suitable stand (as shown in Fig. 48), they were in turn brought immediately under the fixed electrode F (of aluminium or iridio-platinum).

We were soon convinced that it was necessary to regulate with extreme accuracy the length of the spark by which the incandescent vapour of the alloy was produced.

In the first attempts a fragment of the alloy of a more or less irregular shape was simply held in a suitable clip, and a spark-current passed from the alloy to a similar fragment of aluminium or other metal.   We then adopted a method suggested by Professor Stokes, which consists in the passage of the spark-current in a vertical line across the shortest distance between two

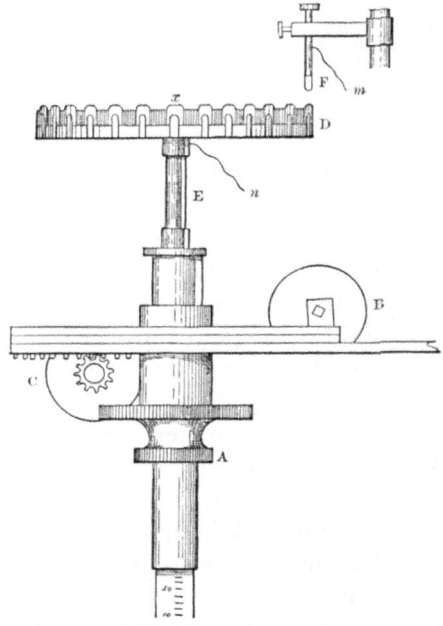

FIG. 48.—SPECIMEN HOLDER.

A. Elevating Screw.--B. Rotating Screw.—C. Screw giving side motion.—D. Wheel holding specimens.—F. Upper fixed pole.—*m n.* Wires to coil.

cylinders arranged in a horizontal position, with their axes at right angles to each other.

The use of these cylinders was attended with many difficulties, and we substituted for them strips of metal cut in the form indicated in the figure (at *x*).

The distance between the poles was at first adjusted by a gauge, and subsequently by a cathetometer; but the most accurate results were obtained by placing the portion of alloy in the field of a microscope furnished with a 3- or 4-inch objective, a simple mechanical arrangement, shown in the figure, bringing the surface from which the spark would pass to the point of intersection of two spider-lines in the eyepiece of the microscope.

We were careful to select alloys which were homogeneous in character; and our attention was first devoted to observations on the zinc-cadmium alloy, one of a series of alloys termed by Matthiessen "solidified solutions of one metal in another."

We next operated upon the gold-copper alloy, as its molecular arrangement appeared to render it peculiarly suited for the purpose of the research, the use of this alloy being attended with the additional advantage that the ordinary method of assay has rendered it possible to determine its composition with accuracy to the $\frac{1}{10000}$ part of the original weight of the assay-piece, a degree of precision which will appear remarkable to all who are familiar with the ordinary methods of quantitative analysis.

The results of the observations were represented by curves, the coordinates being the composition of the alloys as determined by the ordinary method of assay, and the varying points of equal brightness measured by the micrometer of the observing-telescope fitted with movable horizontal wires. The micrometer-readings are represented by spots placed opposite the compositions as determined in the ordinary manner, a mean curve being subsequently constructed.

Our method of procedure was upon each occasion to prepare such a curve by means of accurately known standards; and having this curve, to determine, by means of the micrometer-readings, the positions which various specimens of unknown composition would occupy on it. By carrying the eye from the curve to

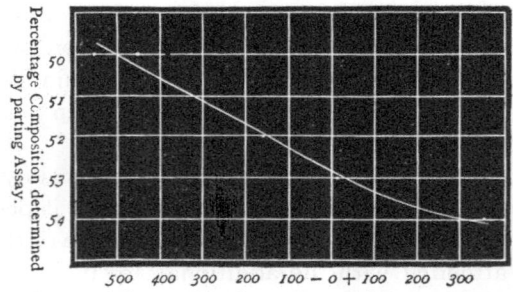

FIG. 49.—Micrometer Readings.   Zinc and Cadmium Alloys.

the side on which the parting assay determinations were shown, we were enabled to compare the places assigned to the specimens on the curve, by the spectroscope, with those determined by the parting assay which we then learnt for the first time. As the result of many series of experiments with alloys of unknown composition, we found that the difference between the spectroscopic and parting assays was but small so long as the conditions under which we experimented continued to be uniform.

The following experiments may be given :—
A number of alloys of zinc and cadmium were syn-

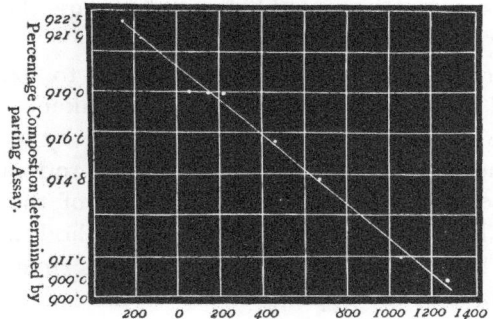

FIG. 50.—Curve A.   Gold and Copper Alloys.

thetically prepared, and from these a series of five, the
percentages of cadmium in which increased by 1 per

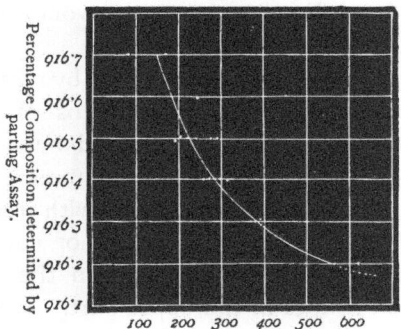

FIG. 51.—Curve B.   Gold and Copper Alloys.

cent. from 50 per cent. to 54 per cent., were selected.
It is interesting to note that both the metals employed

in the alloy are very fusible (zinc melting at 433° C. and cadmium at 228° C.) ; and observations by means of the spectroscope at once enabled us to arrange these alloys in their correct order.   See Fig. 49.

A series of alloys ranging from 900·0 to 922·5 parts of gold in 1000 of the alloy was next taken.   A curve (A) was obtained by measurements made by placing the micrometer-wires in a vertical position and measuring the base of the lowest visible portions of the spectroscopic lines under examination, a method which appeared to be better suited to wide ranges than observations made in the upper portions of the spectral lines ; these latter were, however, found more suited to the study of narrow ranges.   Our attention was mainly devoted to ascertaining whether the curves would be continuous, and at what point a curve deduced from any given lines in the spectrum would cease to be sensitive.

Another curve (B) was constructed by means of observations made upon several specimens of gold-copper alloy, the composition of which had been determined to the $\frac{1}{10000}$ part in the ordinary manner.

It will be seen at a glance that with regard to the lower part of the curve a change of composition of $\frac{1}{10000}$ gives a much greater change in the micrometer-readings than it does in the upper part.   It is also to be remarked that, as in this more sensitive portion a change of $\frac{1}{10000}$ part is represented by a change of 200 in the micrometer-readings, a true mean curve can only be constructed when the actual composition of the alloy is stated with much greater accuracy than that afforded by the present method of assay.

In several cases duplicate and even triplicate readings were made with the same specimens, the number of spots placed opposite any figure in the vertical column does not therefore necessarily indicate the number of pieces of that composition under examination. If we examine the Mint specimens, of which the composition is stated to be 916·4, it is seen that the maximum departure from the curve is less than $\frac{1}{20000}$ part; and in this case it is impossible to say that any error has been made by the spectroscope, because, were the composition of all the alloys in question to be what the spectroscope states them to be, they would still be called 916·4 on the parting assay. Again, if we consider the specimens 916·6, it will be seen that while in those two with micrometer-readings 150 the departure is not more than that previously stated, $\frac{1}{20000}$ part; on the other hand, the composition of the two the micrometer-readings of which were 250, would carry them into a region of the curve which would spectroscopically indicate their composition to be less than 916·5. The departure in this case was so great that it was considered advisable to make a fresh determination of their composition by the ordinary method ; and it was satisfactory to find that the parting assay determination, as revised, was almost identical with that given by the spectroscope.

§ 7. *The Work of the Future.*

It is impossible to foresee to what analytical operations the new method may be found to be applicable ; but as the experiments now in question were mainly directed towards the development of a new method of assaying gold, it is advisable that attention should be limited to the relative advantages of the old and new methods.

We ascertained by careful experiment that the amount of alloy actually volatilized during an observation in no case exceeds 0·0001 grm. ; and it is interesting to compare this with the amount of metal usually employed in assaying gold, which varies from 0·5 to 1 grm.

It may be objected that the amount of metal employed in the new method is very small ; but granting the accuracy of the method and the homogeneity of the alloy— there is the rub—there is of course no reason why the composition of a gold ingot may not be ascertained by it as accurately as by the old method.

It is asserted that the existing method is usually accurate to the $\frac{1}{10000}$ part of the portion of metal operated upon, the process possessing in addition to its accuracy many incidental advantages, not the least of which is the possibility which it affords of detecting, at different stages of the work, the presence of metallic impurities, such as iridium.

On the other hand, it should be observed that the method now in use comprises six distinct operations, and two hours are occupied in ascertaining the final result of the assay ; and as it is frequently important to determine the value of an ingot of gold with rapidity, it will be obvious that the new method possesses marked advantages in this respect, for by its aid a result may be obtained in a few minutes.

Experience alone can show whether this new process may be made as trustworthy as the existing method ; and hitherto there has not been absolute identity in the conditions under which the several experiments were made ; for instance, the intensity of the current employed to volatilize the metal varied from time to time ; this, together with other defects, mainly arose from the fact that the problem has been considered solely from its scientific aspect.

Whether this new method be found preferable to the old one or not, the fact has been established that it is possible to detect, by its means, differences in the composition of the gold-copper alloy as minute as the $\frac{1}{10000}$ part of the whole mass.

I conclude this chapter by expressing my obligations to Dr. Siemens and other friends, who have sent me specimens of various mixtures by which to test the process. I regret that my time has been so much taken up in other ways, that I have been unable to make any experiments beyond those recorded above.

CHAPTER IX.

ON THE COINCIDENCES OF SPECTRAL LINES.

§ 1. *General Presence of Impurities.*

A PROLONGED examination of various spectra is not required to afford evidence not only of the great impurity of most of the metals used, but of the fact that many, if not all, of the coincidences observed by Thalén and others may be explained without having recourse to the idea of physical coincidences.

A study of Ångström's map of the solar spectrum, to take an instance, shows many cases in which a line has been observed to be common to two or more spectra ; and this is especially the case with the lines of iron, titanium, and calcium, nearly every other solar metallic spectrum exhibiting one or more cases of coincidence with the latter.

In those cases which have been examined, it has frequently been found that a line coincident in different spectra is long and bright in only one of them, and that in others it is short, or faint, or both ; or it may even, in certain specimens of the substances, be altogether absent from the spectrum.

As an instance of this difference of behaviour, the

following cases in the spectra of calcium and strontium may be given. The longest line in the visible portion of the calcium spectrum (Thalén) wave-length 4226·3, is found in the strontium spectrum as a line of medium length. 4607·5, one of the longest lines of strontium, appears in the calcium spectrum as a short line.

Another very long line of strontium occurs at 4215·3, in close proximity to the longest calcium line, and, according to Thalén, occurs also in the spectrum of that metal.

I did not at first succeed in obtaining any evidence of its presence in the calcium spectrum ; but the metal I employed was very pure.

We have here, then, two metals with two lines common to their spectra ; and it is found that the line which is long and bright in one spectrum, is faint in the other ; and with regard to a third line, one observer finds it in both spectra, the other in one only, and after many attempts succeeds in observing it in the second, *but only in a specimen known to be contaminated with the first.*

The simplest explanation of the case, bearing in mind the facts already dwelt upon, is that the calcium used to produce the spectrum was contaminated to a certain extent with strontium, the strontium in turn containing calcium—a state of things which a moment's consideration will show to be not only possible but most probable, the close chemical relation of the two metals, and the extreme difficulty of making even an approximate separation when mixed, being well known.

Even if we knew nothing of the probability of mixture occurring in the cases of the two metals in question, the behaviour of the line at w. l. 4215·3 is sufficient to show what is the true cause of the coincidences.

The long lines of calcium at wave-lengths 4226, 3968, and 3933 occur also in the spectrum of iron, cobalt, nickel, barium, and other metals as observed in the arc *using carbon poles,* and assume very considerable proportions, equalling or surpassing in length many undoubted lines of those elements which are less easily volatilized by the actions of the current ; on the other hand, the iron lines at wave-lengths 4071, 4063, and 4045, occur in calcium, strontium, and barium, and in other metals under like conditions.

Again, the longest lines of aluminium (wave-lengths (3961 and 3943) occur usually in the spectrum of iron as longish lines, and are to be found in the spectra of cobalt, nickel, calcium, strontium, and barium, and in other metals, where they are even longer than some of the true lines of the metals in which they occur owing to their lower volatility.

§ 2. *General Statement Concerning Coincidences.*

As a result of these considerations the following general statements may be hazarded, premising that it is possible that further inquiry may modify them.

1st. If the coincident lines of the metals are considered, those cases are rare in which the lines are of the first order of length in all the spectra to which they are common, especially if the volatility of the metals in question is about the same : Those cases are much more common in which they are long in one spectrum and shorter in the others.

2nd. As a rule, in the instances of those lines of iron, cobalt, nickel, chromium, and manganese, substances volatilized with difficulty, which are coincident with lines of calcium (which volatilizes easily), the calcium lines are long. Hence we are justified in assuming that lines of iron, cobalt, nickel, chromium, and manganese, coincident with long and strong lines of calcium, are really due to traces of the latter metal occurring in the former as an impurity.

3rd. In cases of coincidences of lines found between the lines of various spectra, the line may be fairly assumed to belong to that one in which it is longest and brightest.

In order to show what a fair promise there is that all these questions will in time be set at rest by photography, and set at rest in the direction I have indicated, I here reproduce one of the very earliest photographs taken by the method which has been described previously. It is a confronting of the spectra of the metals calcium and strontium.

A simple comparison of the two spectra shows that there are three strong and thick lines common to the two at w. l. 3968, 3943, the two H lines, and 4226·3, the large calcium line near G. The spectrum of strontium shows three lines which are very much thicker than the corresponding lines in the (calcium) spectrum ; they are situated near wave-lengths 4029 and 4077 and at 4215·3, The latter line has been ascribed by Thalén to calcium, and is coincident with a strong solar line. An inspection of the photograph, however, at once showed me that this line is really a strontium line, since it is thickest in the spectrum of that metal ; so that this single plate was at once sufficient, in the light of these researches, to establish the presence of strontium in the reversing layer of the sun.

It is seen, then, that a comparison of a photograph of any spectrum with the photographs of the other spectra in which coincident lines occur, will be sufficient to show to which spectrum a disputed line belongs. It will be also noticed that the three calcium lines first mentioned are nearly as thick in the lower (strontium) spectrum as in that of calcium itself, while the difference between the thick lines of strontium and the corresponding lines also visible in the calcium spectrum is very great. All these facts are easily explained on the supposition that the calcium was very much less impregnated with strontium than the strontium with calcium. In fact I had such faith in the efficacy of the method, and in the opinion that coincidences are merely due to impurities, that I did not even consider it necessary to change the poles, but proceeded at once to place the strontium salt on that which had just before served for the ignition of the calcium.

This at once accounts for the greater impurity of the former. In nearly pure strontium the same lines are seen, but they are then thinner and shorter.

I may add that the lines at 4045, 4063, and 4071 are due to an iron impurity ; these are the longest lines of iron in that portion of the spectrum photographed.

§ 3. *Elimination of Impurity Lines.*

These considerations therefore supply us with a method of *elimination of lines due to impurities* from maps of the spectrum of metals. The process is conducted as follows :—The spectrum of the element is first confronted with the spectra of the substances most likely to be present as impurities, and with those of metals which, according to Thalén's measurements, contain in their spectra coincident lines.

Lines due to impurities, if any are thus traced, are marked for omission from the map and their true sources recorded, while any line that is observed to vary in length and thickness in the various photographs is at once suspected to be an impurity line, and if traced to such is likewise marked for omission.

The retention or rejection of lines coincident in two or more spectra is determined by observing in which spectrum the line is thickest ; where several elements are mapped at once, all their spectra are confronted on the same Plate, as by this means the presence of one of the substances as an impurity in the others can be at once detected.

Thus the two lines H and K (3933 and 3968), assigned both to iron and calcium by Ångström, are proved to belong to calcium by the following observations :—

a. The lines are well represented in the spectrum of commercial wrought iron, but are absolutely coincident with two thick lines in the spectrum of calcium chloride with which the iron spectrum has been confronted.

b. The lines are represented by mere traces in the spectrum of a specimen of pure iron prepared by the late Dr. Matthiessen, and obligingly placed at my disposal by Dr. Russell. Both poles of the lamp were of iron, the lower pole consisting of an ingot of the metal which had been cast in a lime-mould.

c. The lines are altogether absent in a photograph of pure iron, where both poles of the lamp were of the pure metal *not* cast in lime, and they are likewise absent in a photograph of the spectrum of the Lenarto meteorite.

These examples serve to illustrate the manner in which large numbers of the coincidences recorded by former observers have been disposed of.

In determining the coincidence of very thick lines, such as the H lines just mentioned, the centre of the thick line is taken. It not unfrequently happens that a very thick line will reverse itself, a circumstance which greatly facilitates its comparison with confronted lines, since a thin dark line then runs down the centre of the thicker bright one.

By eliminating lines due to impurities in the manner just described, a spectrum is at length obtained, of which every line is assignable to the particular element photographed.

As the work of mapping progresses, there seems a probability that the spectra of iron and other metals,

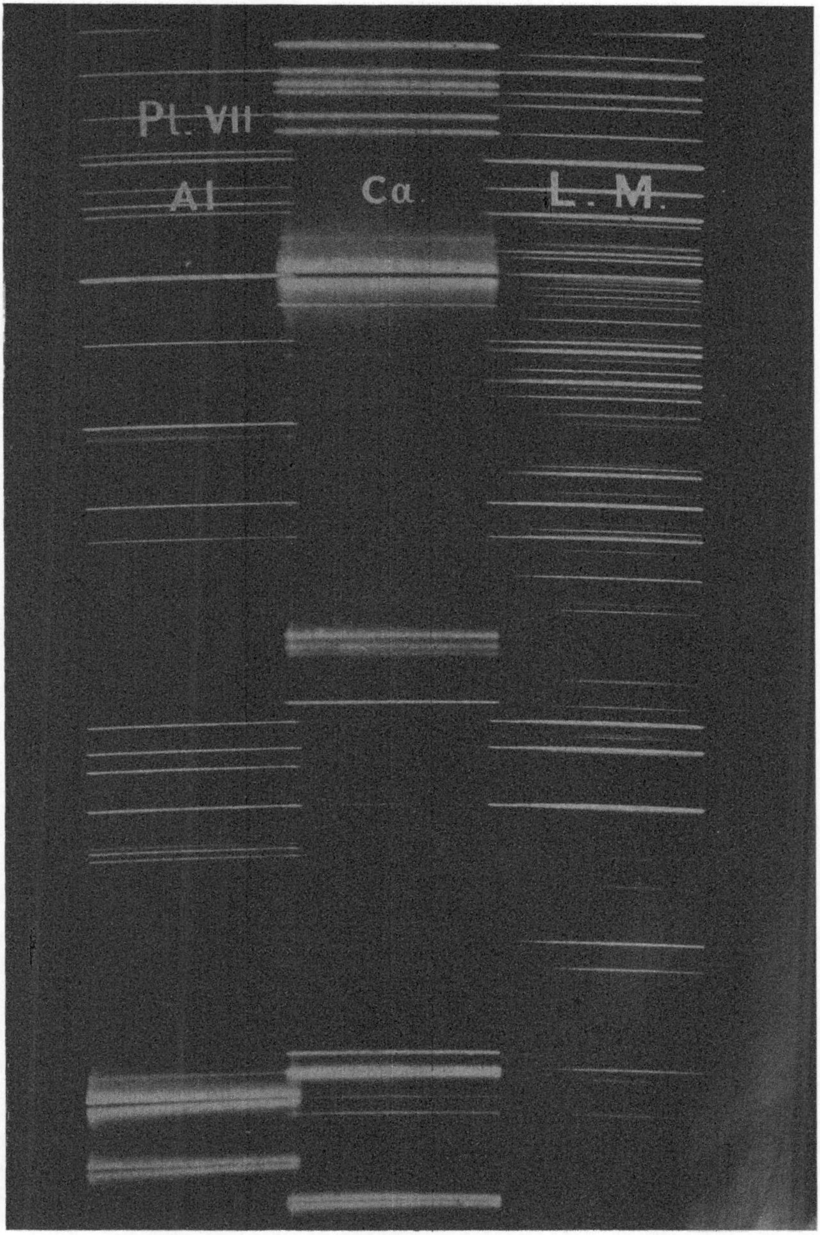

Pl. VII

Al   Ca   L. M.

which all observers have found very complex, will become much simplified, owing to the elimination of many lines hitherto attributed to these metals, but subsequently proving to be due to the presence of impurities in them. It may not be uninteresting to detail a few considerations which induce me to form such a conclusion. Instances have occurred in which a well-defined line appearing in the spectrum of a metal has proved coincident with the longest line in the spectrum of an element newly mapped.

Then, again, suppose iron to be present as an impurity in an element which is being mapped for the first time, the longest lines of iron are first looked for, but it may happen that one of these lines is represented in the spectrum of the new element as very much longer than the other lines in the spectrum of iron, which have hitherto been regarded of about the same length.

Lastly, there are many lines in the spectrum of one element reversed in the solar spectrum, which are absolutely coincident with lines of the same relative wave-length and definition in the spectra of other elements. It may be that greater dispersion may in all these instances prove that these lines, instead of being absolutely coincident, may slightly graze one another ; but my experience up to the present time leads me to suspect that these lines are due to the presence of common impurities, either in the shape of unmapped elements or of elements hitherto unknown.

CHAPTER X.

THE ELEMENTS PRESENT IN THE SUN.

THE observations referred to in Chapter V. have an important bearing upon the study of the chemical constitution of the sun for the reason that, as is well known, all the lines known to exist in the spectrum of an element supposed to be present in the sun's atmosphere are not in all cases reversed.

Before I proceed to give the facts in detail it will be well to go over the prior work of Kirchhoff and Ångström, to see precisely the evidence on which our present knowledge of the elements in the solar atmosphere, as determined by Kirchhoff's method of solar observation (that is, the non-localization or integration of the various solar regions, such as spots, faculæ and chromosphere), rests.

§ 1. *Kirchhoff's Work.*

Kirchhoff, in his Paper referring to Fraunhofer's prior determination of the double line D being coincident with a double line observed in the spectrum of sodium vapour, locates sodium vapour in the solar atmosphere, as

Professor Stokes had done before him. Coincident with all the bright iron lines which he observed with the spark he used (he only saw a small number of the lines), he found well-defined Fraunhofer lines. He therefore located iron vapour in the atmosphere. The rest of the evidence relating to other metals I give from the translation of his memoir by Professor Roscoe.*

" As soon as the presence of one terrestrial element in the solar atmosphere was thus determined, and thereby the existence of a large number of Fraunhofer lines explained, it seemed reasonable to suppose that other terrestrial bodies occur there, and that, by exerting their absorptive power, thay may cause the production of other Fraunhofer lines ; for it is very probable that elementary bodies which occur in large quantities on the earth, and are likewise distinguished by special bright lines in their spectra, will, like iron, be visible in the solar atmosphere. This is found to be the case with calcium, magnesium, and sodium. The number of the bright lines in the spectrum of each of these metals is indeed small ; but those lines, as well as the dark ones in the solar spectrum with which they coincide, are so uncommonly distinct that the coincidence can be observed with very great accuracy.

" In addition to this, the circumstance that these lines occur in groups renders the observation of the coincidence of these spectra more exact than is the case with those composed of single lines. The lines produced by chromium also form a very characteristic group, which likewise coincides with a remarkable group of Fraunhofer lines ; hence I believe that I am justified in affirming the presence of chromium in the solar atmosphere. It appeared of great interest to determine whether the solar atmosphere contains nickel and cobalt, elements which invariably accompany iron in meteoric masses. The spectra of these metals, like that of

---

* Transactions of Berlin Academy, 1861. Translated by Roscoe. Macmillan.

iron, are distinguished by the large number of their lines ; but the lines of nickel, and still more those of cobalt, are much less bright than the iron lines, and I was therefore unable to observe their position with the same accuracy with which I determined the position of the iron lines. All the *brighter lines* of nickel appear to coincide with dark solar lines ; the same was observed with respect to some of the cobalt lines,* *but was not seen to be the case with other equally bright lines of this metal.* From my observations I consider that I am entitled to conclude that nickel is visible in the solar atmosphere ; I do not, however, yet express an opinion as to the presence of cobalt. Barium, copper, and zinc appear to be present in the solar atmosphere, but only in small quantities ; the brightest of the lines of these metals correspond to distinct lines in the solar spectrum, but the weaker lines are not noticeable. The remaining metals which I have examined, viz., gold, silver, mercury, aluminium, cadmium, tin, lead, antimony, arsenic, strontium, and lithium, are, according to my observations, not visible in the solar atmosphere. Through the kindness of M. Grandeau, of Paris, I obtained several pieces of fused silicium ; I was thus enabled, by using them as electrodes, to examine the spectrum of this element. The lines in the silicium spectrum are, however, with the exception of two broad green bands at 1810 and 1830, so deficient in luminosity that I was unable to determine their position with sufficient accuracy to reproduce them in my drawing. The two bright green bands do not correspond to dark bands in the solar spectrum ; so that, as far as I have been able to determine, silicium is not visible in the solar atmosphere."

It will be seen from the foregoing that Kirchhoff deals mainly with the brightest lines, although the test failed him in the case of cobalt, for a reason I shall show further on. Hence, as a result of Kirchhoff's work, we have in the solar atmosphere :—

* The italics are mine.

Present.	Doubtful.	Absent.
Sodium.	Cobalt.	Gold.
Iron.		Silver.
Calcium.		Mercury.
Magnesium.		Aluminium.
Nickel.		Cadmium.
Barium.		Tin.
Copper.		Lead.
Zinc.		Antimony.
		Arsenic.
		Strontium.
		Lithium.
		Silicium.

§ 2. *Ångström and Thalén.*

Ångström* gives no list such as this, but in its place a table of coincidences observed. Thalén, his associate, in a separate memoir,† gives, however, as present in the Sun :—

Sodium,	Chromium,	Hydrogen,
Iron,	Nickel,	Manganese,
Calcium,	Cobalt,	Titanium,
Magnesium,		

thus rejecting zinc and barium from Kirchhoff's list of accepted elements, adding cobalt from the doubtful list, and hydrogen and manganese from Ångström's, and titanium from his own observations.

* Recherches sur le spectre solaire, par A. J. Angström. Spectre normal du Soleil. Berlin, 1869.
† Longueurs d'onde des raies métalliques, p. 11. Nova Acta. Upsala, 1868.

The table of coincidences referred to and Ångström's remarks thereon explain the cause of this. Kirchhoff's evidence for zinc had depended upon the coincidence of two lines only, and these were doubtless thought insufficient, as in the cases of the metals retained in the list the number of the coincidences was much greater, viz. :—

Sodium . . .	9 (all)	Magnesium . .	4 (3?)
Iron	450	Chromium . .	18
Calcium . . .	75	Nickel . . .	33
Cobalt	19	Hydrogen . .	4 (all)
Manganese . .	57	Titanium . .	118
Barium . . .	11 (of 26)	Zinc	2 ? (of 27)
Aluminium . .	2 ? (of 14)		

From Ångström's remarks, which I proceed to give, it is evident that he was not quite satisfied with the brilliancy test relied on by Kirchhoff, and that his doubts concerning zinc arose from this cause.

" L'aluminium possède certainement des raies brillantes en plusieurs endroits du spectre, mais les raies situées entre les deux H sont les seuls qui semblent coïncider avec les lignes Fraunhofériennes. Pour expliquer ce phénomène singulier il faut dire que les raies violettes se présentent comme les plus fortes dans le spectre de ce métal. De même que les raies jaunes du sodium, ces deux raies d'aluminium ont fait voir quelquefois le phénomène d'absorption consistant en ce qu'une raie noire se présente dans le milieu de chacune d'elles, ce qui prouve la forte intensité des dites raies. En observant les rayons extra-violettes de ce métal, on décidera si les deux raies mentionnées ci-dessus coïncident ou non avec des raies Fraunhofériennes ; car si ma supposition est vraie,

Pl. VIII.   SHOWING

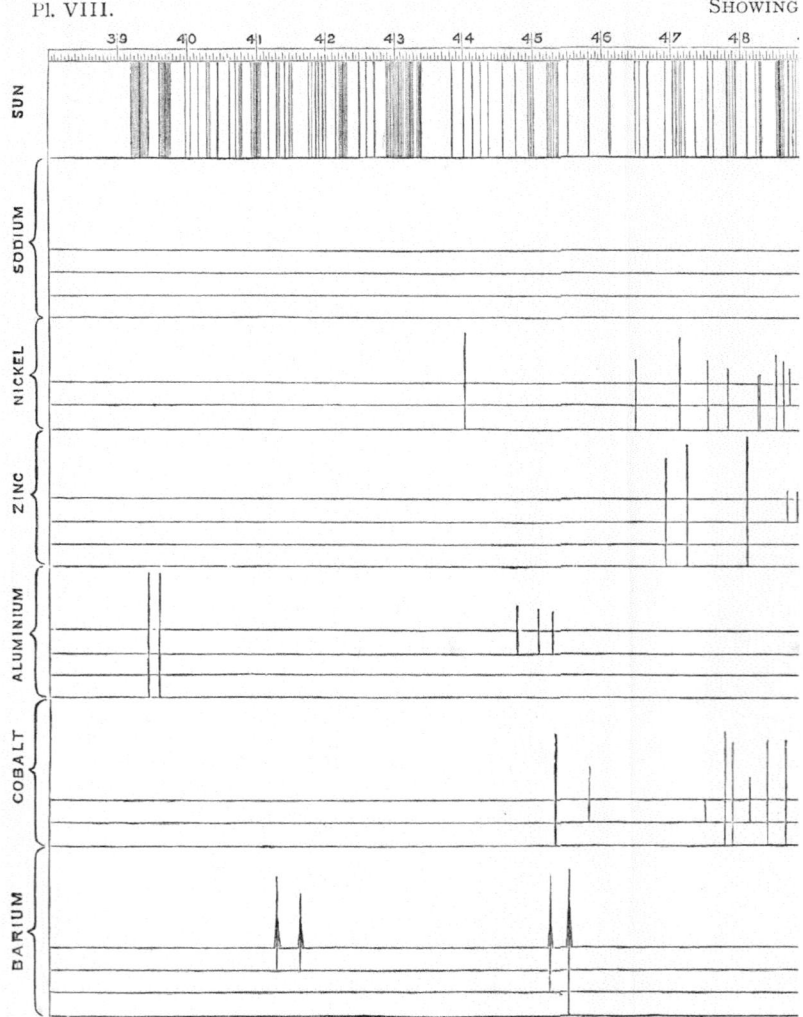

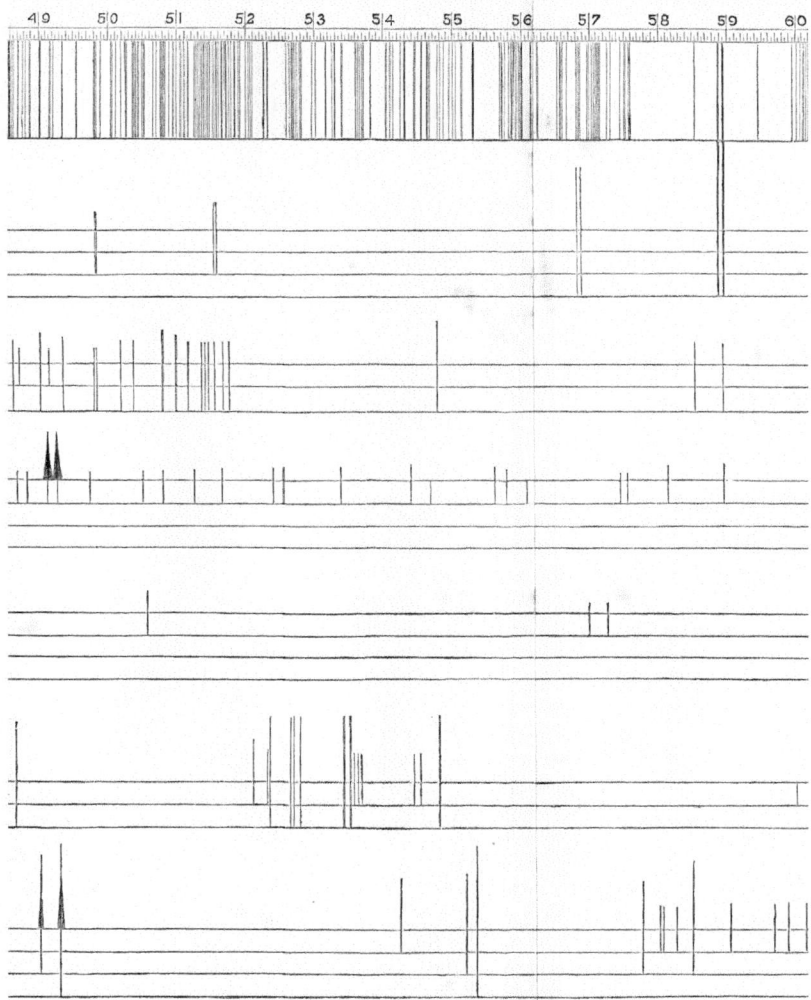

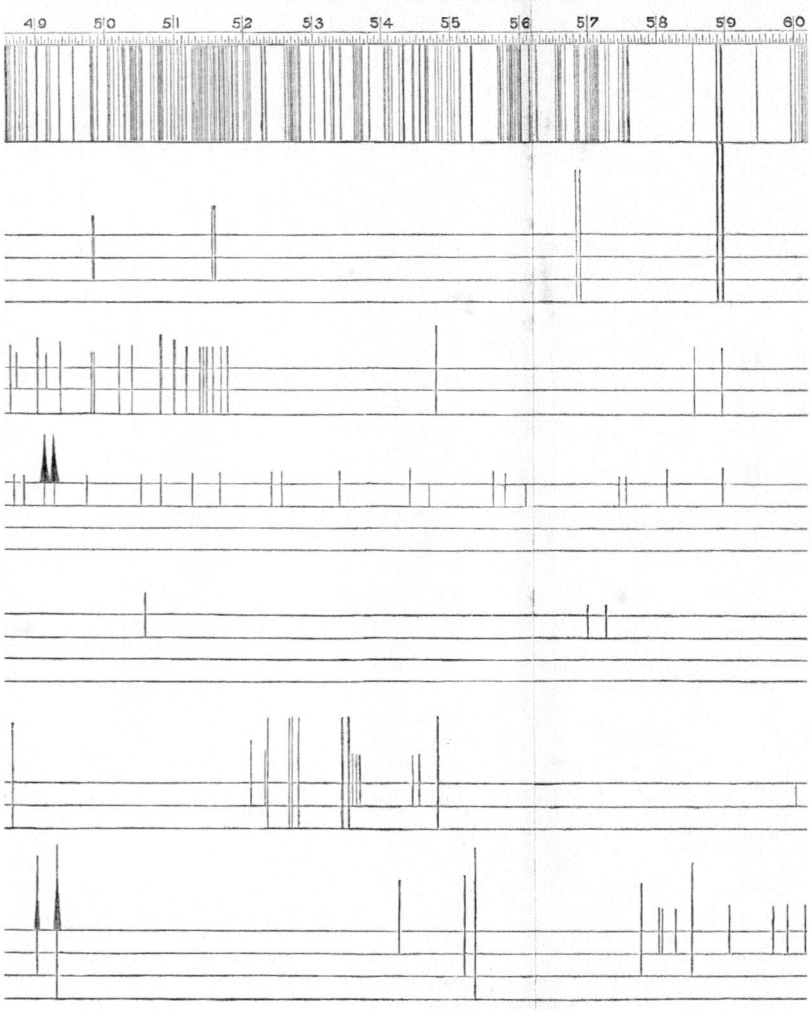

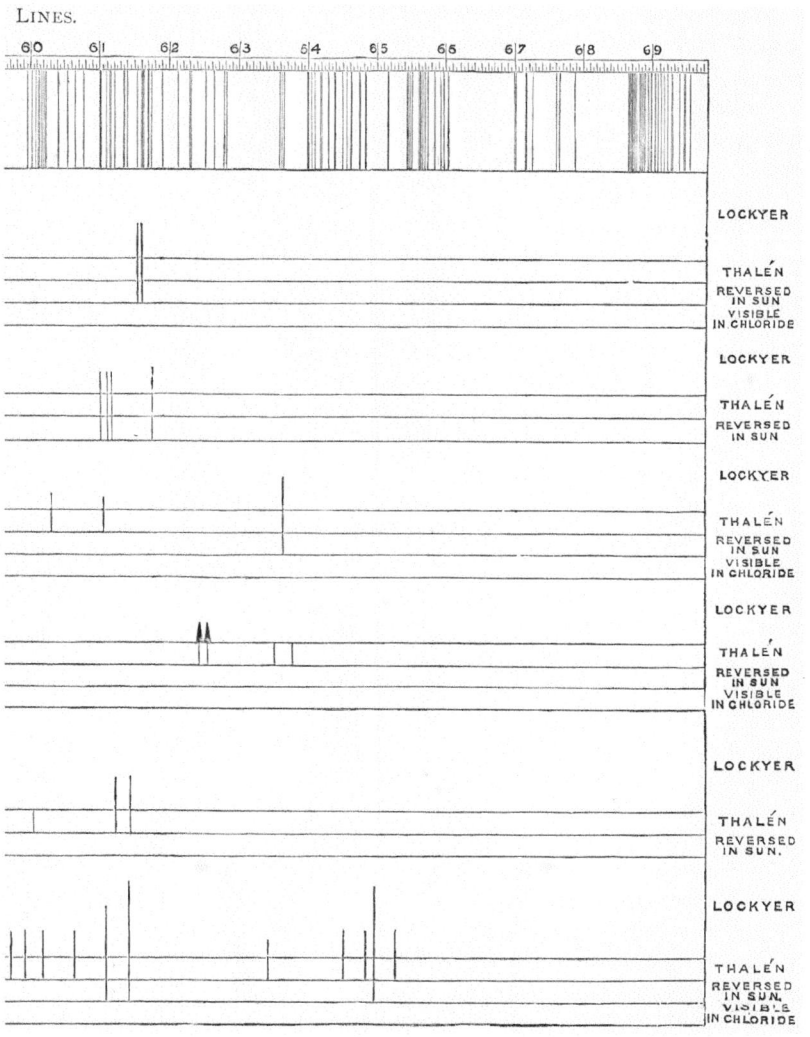

LINES.

6|0   6|1   6|2   6|3   5|4   6|5   6|6   6|7   6|8   6|9

LOCKYER

THALÉN
REVERSED
IN SUN
VISIBLE
IN CHLORIDE

LOCKYER

THALÉN
REVERSED
IN SUN

LOCKYER

THALÉN
REVERSED
IN SUN
VISIBLE
IN CHLORIDE

LOCKYER

THALÉN
REVERSED
IN SUN
VISIBLE
IN CHLORIDE

LOCKYER

THALÉN
REVERSED
IN SUN.

LOCKYER

THALÉN
REVERSED
IN SUN.
VISIBLE
IN CHLORIDE

les raies extra-violettes doivent coïncider aussi avec les lignes obscures du spectre solaire.

" A deux raies du zinc que j'ai indiquées sur mes planches comme coïncidentes avec des raies Fraunhofériennes il en faut ajouter une troisième, située à 4809·7 ; mais, à l'égard des deux raies, très-larges et très-fortes, d'une apparence nébuleuse, il n'y a pas de correspondance visible ; ainsi, la présence du zinc dans le soleil me semble très-douteuse. Je dirai cependant qu'il existe trois raies de magnesium, du même aspect nébuleux, qui ne possèdent pas non plus de correspondance avec les raies de Fraunhofer, quoique la présence de ce corps dans le soleil ne permettre pas le moindre doute."

3. *Study of Lines actually Reversed.*

Now the point of the recent work is that when we compare the spectra of metallic vapours reversed in the solar spectrum, such spectra being mapped by the new method, and showing the long and short lines, *the reversed lines are invariably those which are longest in the spectrum.* Here we have in fact the true test to apply to the reversal of solar lines, and a guide of the highest value in spectrum observations of the chromosphere and·photosphere. It was seen at once that to the last published table of solar elements (that of Thalén) must be added zinc, aluminium, and possibly strontium as a result of the application of the new test.

In order to pursue the inquiry under the best conditions, complete maps of the long and short lines of all the elements are necessary. But it was not absolutely necessary for the purposes of a *preliminary* inquiry to wait for such a complete set of maps, for the lists of lines

R

given by the various observers may be made to serve as a means of differentiating between the longest and shortest lines, because as we have seen the lines given at a low temperature, by a feeble percentage composition, or by a chemical combination of the vapour to be observed, are precisely those lines which appear longest when the complete spectrum of the pure dense vapour is studied.

Now with regard to the various lists and maps published by various observers, it is known (1) that very different temperatures were employed to produce the spectra, some investigators using the electric arc with great battery-power, others the induction-spark with and without the jar ; (2) that some observers employed in certain cases the chlorides of the metals the spectra of which they were investigating, others used specimens of the metals themselves.

It is obvious, then, that these differences of method could not fail to produce differences of result ; and accordingly, in referring to various maps and tables of spectra, we find that some include large numbers of lines omitted by others. A reference to these tables, in connection with the methods employed, shows at once that the large lists are those of observers using great battery-power or metallic electrodes, the small ones those of observers using small battery-power or the chlorides. If the lists of the latter class of observers be taken, we shall have only the longest lines, while those omitted by them and given by the former class will be the shortest lines.

In cases, therefore, in which I had not mapped the spectrum by the new method of observation referred to in my paper, I in the first instance took the longest lines as thus approximately determined ; for it seemed desirable, in view of the very large number of unnamed lines, to search at once for the longest elemental lines in the solar spectrum without waiting for a complete set of maps.

§ 4. *Result of Preliminary Search.*

A preliminary search having been determined on, I endeavoured to get some guidance by seeing if there was any quality which differentiated the elements already traced in the sun from those not traced ; lists were prepared showing broadly the chief chemical characteristics of the elements traced and not traced. This was done by taking a number of the best known compounds of each element (such, for instance, as those formed with oxygen, sulphur, chlorine, bromine, or hydrogen), stating after each whether the compounds in question were unstable or stable. Where any compound was known not to exist, that fact was indicated.

Two Tables were thus obtained, one containing the solar, the other the more important non-solar elements (according to our knowledge at the time).

These Tables gave me, as the differentiation sought, the fact that in the main the known solar elements formed stable oxygen-compounds. I have said in the main, because the differentiation was not absolute ; but

R 2

it was sufficiently strong to make me commence operations by searching for the outstanding strong oxide-forming elements in the sun.

The immediate result was that *strontium, cadmium, lead, copper, cerium,* and *uranium* were shown with considerable probability to exist in the solar reversing layer, in addition to the elements in Thalén's last list.

As another test, certain of those elements which form unstable compounds with oxygen were also sought for, gold, silver, mercury being examples. None of these were found.

This was in 1874. The result up to that time, then, was as follows :—

METALS PRESENT IN THE SUN.

Sodium	Cobalt	Cadmium
Iron	Hydrogen	Lead
Calcium	Manganese	Copper
Magnesium	Titanium	Cerium
Chromium	Strontium	Uranium
Nickel		

§ 5. *Subsequent Inquiry.*

I have been prosecuting a much more rigid inquiry since that date, as I have been making efforts to construct a new map of the solar spectrum. In the course of this work the spectra of most of the metallic elements have been directly compared by the method before stated in Chapter III. The following Tables give the present state of the inquiry (Nov., 1877).

The lines given in evidence are the longest visible in the photographic region of the respective spectra, so that the fact of their being reversed in the solar spectrum must be considered as strong evidence in favour of the existence of the metals to which they belong in the sun. Where, however, there is only one line, as with Lithium, Rubidium, &c., the evidence cannot be considered final, and until a larger number of coincidences is determined, the presence of these metals in the sun's reversing layer can only be said to be probable.

It must be borne in mind that, in addition to the long lines which a spectrum may contain in the red, yellow, or orange, long lines may exist in the hitherto unexplored ultra-violet region, so that the absence of such metals from the sun cannot be absolutely affirmed until a complete survey of these regions has been completed. This is a work of time.

TABLE I.

ELEMENTS WHOSE PRESENCE IN THE SUN'S REVERSING LAYER HAS BEEN CONFIRMED.

Name of Element.	Evidence.	Authority.
Sodium	Reversal of D lines.	* S. K.
Iron	Reversal of 450 lines.	K.
Calcium	Reversal of 75 lines.	K.
Magnesium	Reversal of 4 (3 ?) lines.	K.
Chromium	Reversal of 18 lines.	K.
Nickel	Reversal of 33 lines.	K.
Barium	Reversal of 11 lines (of 26).	K.

* S. = Stokes. K. = Kirchhoff. A. = Ångström. T. = Thalén. L. = Lockyer.

Name of Element.	Evidence.	Authority.
Zinc *	Reversal of 2? lines (of 27).	K.
Cobalt	Reversal of 19 lines.	T.
Hydrogen.	Reversal of 4 lines (all),	A.
Manganese	Reversal of 57 lines.	A.
Titanium	Reversal of 118 lines.	T.
Aluminium	Reversal of the 2 longest lines at w. ls. 3943·30 and 3960·50.	L.
Strontium	Reversal of 4 lines at w. ls. 4029·6, 4076·77, 4215·00, and 4607·5.	L.
Lead	Reversal of 3 lines at w. ls. 4019·28, 4056·8, and 4061·25.	L.
Cadmium	Reversal of 2 lines at w. ls. 4677·0 and 4799·00.	L.
Cerium	Reversal of 2 lines at w. ls. 3928·7, and 4012·0.	L.
Uranium.	Reversal of 3 lines at w. ls. 3931·0, 3943·0, and 3965·8.	L.
Potassium	Reversal of 2 lines at w. ls. 4042·75, and 4046·28 (apparently the only K lines in this region of the spectrum).	L.
Vanadium	Reversal of 4 lines at w. ls. 4379·0, 4384·0, 4389·0, and 4407·5.	L.
Palladium	Reversal of 5 lines at w. ls. 3893·0. 3958·0, 4787·0, and 4817·0, and 4874·0.	L.
Molybdenum	Reversal of 4 lines at w. ls. 3902·0, 4576·0, 4706·0, and 4730·0.	L.

* Thalén excluded this metal from the list of solar elements subsequently to Kirchhoff's including it. My observations confirm those of Kirchhoff.

TABLE II.

ELEMENTS WHOSE PRESENCE IN THE SUN'S REVERSING
LAYER IS PROBABLE.

Name of Element.	Reason of Doubt.	Authority.
Indium	One line at w. l. 4101·0 is apparently coincident with h, hitherto regarded as a hydrogen line. The reversal of another line at w. l. 4509·0 is doubtful.	L.
Lithium	One line at w. l. 4603·0 is reversed, but the reversal of the long red line at w. l. 6705 has not yet been detected.	L.
Rubidium	One long line at w. l. 4202·0 is reversed, but solar lines corresponding to the long red lines at w. ls. 6205 and 6296 have hitherto escaped detection.	L.
Cæsium	Two lines at w. ls. 4554·9 and 4592 are possibly reversed, but a better photograph is needed to settle the question.	L.
Bismuth	One line at w. l. 4722·0 is reversed, but further evidence considered necessary.	L.
Tin	One line at w. l. 4524 is apparently reversed, but further evidence is desirable.	L.
Silver	Two lines at w. ls. 4018·0 and 4212·0, which are reversed in the metallic spectrum, are of very great width, and I have not yet had time to determine whether they are coincident or not with lines in the solar spectrum, by alloying silver with copper or some other metal so as to thin down the lines.	L.
Glucinum	One line at w. l. 3904·77 is apparently reversed, but further evidence desirable.	L.
Lanthanum	Three winged lines at w. l. 3948.20, 3988.0, and 3995.04 are reversed.	L.
Yttrium or Erbium	Two lines at w. l. 3981·87 and 3949·55 are reversed.	L.

TABLE III.

ELEMENTS ABSENT FROM SUN'S REVERSING LAYER, SO
FAR AS OUR KNOWLEDGE AT PRESENT EXTENDS.

Name of Element.	Evidence.	Authority.
Carbon	No coincidences with the carbon lines.	A.
Silicium	No reversals determined.	K.
Thallium	The long green line at w. l. 5349 is apparently not reversed.	L.
Chlorine Bromine Iodine	No coincidence observed between solar lines and the bright lines seen in the jar-spark spectrum.	L.

The results given in this last table have a special
interest, because I have grounds for thinking that in
the case of carbon and iodine there are some coincidences with the most marked parts of their *fluted
spectra ;* this may be presumed to indicate that in a
region of the sun's atmosphere cooler and therefore
higher than the reversing layer these elements may
exist in a state of greater molecular complexity than
do the metallic vapours lower down.

§ 6. *Mitscherlich's Observations.*

This enables us to discuss what Mitscherlich, in his
first memoir, states as to the bearing of . his observations with regard to the solar spectrum :—

" Ces essais montrent comment l'analyse spectrale peut conduire à la connaissance des affinités mutuelles des corps simples à la température de l'atmosphère solaire. Si l'on observait, par exemple, le spectre d'un chlorure alcalin terreux dans la lumière du soleil, on en pourrait conclure que son métal possède, à la température du soleil, une affinité pour le chlore plus grande que le potasium ou le sodium, ces derniers métaux existant à l'état de liberté. Reciproquement, la connaissance des combinaisons existant dans l'atmosphère solaire pourra conduire à connaître la température de cet astre, si toutefois nous ne parvenons jamais nous-mêmes à approcher de cette température.

" La présence du sodium libre dans l'atmosphère conduit à admettre qu'il n'y existent pas de corps electro-négatifs libres, tels que l'oxygène ou le soufre, et qu'ils n'y existent même pas autrement en quantité assez abondante pour se combiner avec tout le sodium. En outre, tous les métaux que le sodium chasse de leurs combinaisons doivent aussi y exister à l'état de liberté.

" Les nouveaux spectres que j'ai fait connaître pourront conduire à constater dans l'atmosphère solaire la présence du chlore, du brome, de l'iode, du phosphore, etc.

" D'un autre côté, de l'absence des raies d'un métal dans le spectre solaire on ne saurait conclure celle du métal lui-même dans l'atmosphère du soleil ; il peut, en effet, s'y trouver des métaux, le lithium, par exemple, qui y sont engagés dans des combinaisons qui ne donnent pas de spectre."

In what has gone before there is, I think, ample evidence that the explanation advanced by Mitscherlich is absolutely untenable ; for at the temperature of the sun, which is high enough to allow hydrogen and even sodium to exist uncombined and in a state of incandescence above the photosphere, there would be heat enough to dissociate compounds, and therefore to cause the longest lines, at all events, of the metalloids to be visible even if they existed in combination as a rule, but, as we have seen, no trace of any line spectrum of the metalloids in the solar spectrum has been recorded.*

As further evidence that there is no chemical combination whatever in the photosphere, the structure of the spectrum may be also instanced ; it certainly would be very different from what it is, did compounds exist in the solar atmosphere ; the least refrangible end of the spectrum would, I hold, be the more, instead of the less complex ; and although Professor Young has recently recorded in the spectrum of a sun-spot certain appearances which might be imagined to favour the idea of the existence of compounds in the comparatively cold down rush into a spot, the general facts, to say the least, seem to point the other way, and in all my observations of sun-spots I have never seen anything approaching to the appearance put on by a compound spectrum. Still I am far from committing myself on this point, and am waiting for a sun-spot to determine the question by means of photography.

I would further remark, that with our present knowledge it is not difficult to gather from Father Secchi's observations on stellar spectra, that if the atmosphere of a star contains compound molecules, they at once make themselves very obviously visible. Several stars, the spectra of which have been mapped by him, have undoubtedly, in my opinion, atmospheres containing compound molecules ; and it may be that the phenomena of variable stars may be connected with a delicate

---

* *See* also Ångström, ' Recherches sur le Spectre Solaire,' p. 37.

state of equilibrium in the temperature, so that at one time we get the feeble line absorption of the *dissociated*, and at another the strong band-absorption of the *associated* elements in their atmospheres.   In this conclusion, drawn from my long and short line observations, I was anticipated by Angström, who reasoned from less precise data.   Father Secchi's idea, that we have in such stars a prevalence of spot-spectrum,* will, I think, not hold ; but this point, which is one of extreme interest, is still *sub judice*.

* Secchi, 'Le Soleil,' p. 288 *et seq.*

INDEX.

———•+——

A.

Absorption phenomena, 38, 66, 75, 136
—— various kinds of, 67
Air, compression and rarefaction of, 15
—— particles, motion of, 12
Aluminium, its relation to spectra of other metals, 230
Amalgams, experiments, 211
—— method employed in experimenting on, 212
—— substances used in experimenting on, 213
Ångström, 135, 147
Assaying gold, relative advantages of old and new methods of, 226
Atmosphere, density of, 24
—— sound in relation to, 23
Atom, definition of an, 113
Atoms of chemical substances, fundamental lines of, 206

B.

Barium, experiments on, 156

Binary compounds, spectra of, 189
Bunsen, observations of Professor, 178, 200, 201
—— burner, its use for showing the action of heat upon salts, 57

C.

Cadmium, spectrum of, 146, 217
Calcium, blue end of spectrum of, 191
—— dissociation of, 197
—— experiments on blue end of spectrum of, 103
—— its relation to spectra of other metals, 230
Calcium and Strontium, comparison of lines of, 229, 231
—— chemical relations of, 229
Chemical elements, spectra of, 182
—— compound nature of, 189
Chlorides and metals, difference of spectra of, 152
Chromosphere, observations on, 147
Clark Maxwell, Professor, 114, 117, 135
Clifford, Professor, illustration of atoms by, 115

Woodfall & Kinder, Printers. Milford Lane, Strand. London, W.C.